Earthling

A New Ethics for the Anthropocene

By

Dean Wallraff

Earthling: A New Ethics for the Anthropocene

By Dean Wallraff

This book first published 2023

Ethics International Press Ltd, UK

British Library Cataloguing in Publication Data

A catalogue record for this book is available from the British Library

Print Book ISBN: 978-1-871891-67-6

eBook ISBN: 978-1-871891-68-3

Dedicated to the memory of Charles F. Wallraff, 1909-1991

TABLE OF CONTENTS

TABLE OF FIGURES

Preface

Why I'm writing this book

This book grew out of a set of essays I started in 1990, to be collected into a book that would be called Millennium. I was planning to finish it by the year 2000. The main theme was that we should take a longer-term approach—a thousand-year vista—to many of the issues facing us today. The book was written party as a memorial to my father, Charles F. Wallraff, a philosophy professor at the University of Arizona, who died in 1991.

I grew up in the sixties. I was twelve years old when JFK was assassinated in 1963, and halfway through college when the decade ended in 1970. I was a few years too young to synch with the main wave of anti-war protests and hippiedom, but I was heavily influenced with the idea that my generation—baby boomers—would make the world fairer and more egalitarian. It deeply disappoints me, and feels like a personal failure, that this has not come to pass.

I've been a hiker and a lover of the outdoors, especially mountains, since I got married in 1975. My father used to talk about hiking with his father in Yosemite, and near San Luis Obispo, where he grew up. But he never hiked in Tucson, where we lived. It's a city surrounded by four gorgeous mountain ranges, with wonderful hiking. But, strange to say, he never seemed to get the urge to visit them. I started hiking seriously in those mountains after my wife and I moved to Boston, when we came back for a couple weeks every Christmas to visit our families. I hiked a lot more after we moved to California in 1990.

I started writing this preface in the summer of 2015 at Glen Aulin in Yosemite National Park. I was staying in the High Sierra Camp next to the Tuolumne River, and sitting on a bench with a view of a pool and small waterfall on that river. I'd just hiked in from Tuolumne Meadows earlier in the day. I was a member of the Sierra Club Board of Directors and the hike reminded me of the Club's deep connection to Yosemite. Until the 1960s, the focus of the environmental movement was the preservation of iconic public lands. Appreciating the beauty of nature led to a desire to preserve it. This perspective is called elitist now, because supposedly only rich, white men had the money and leisure to visit national parks. But I think most earthlings, even poor people who live in cities, have some opportunity to see the beauty of nature. The resulting love for this planet we live on can be a powerful force for preserving it.

We have more practical reasons now to deal with the most pressing environmental problem in history, namely the climate crisis, which will harm most of the plants and animals on the planet, and dramatically change many ecosystems, including iconic ones like Yosemite.

The environmental movement has lost its way. It tries to fight global warming, but hasn't figured out how to do that effectively. And it's dissipating its strength in a sea of political correctness, focusing too much on social-justice issues.

A decade and a half ago I made my living as a software architect working primarily in the payments area for big New York banks. At the same time, I was the Chapter Conservation Chair for the Angeles Chapter of the Sierra Club. This was a volunteer position overseeing the thirty-five environmental campaigns the Club was running in Los Angeles and Orange Counties. I decided that I didn't want my life to go toward making the wheels of commerce function a bit more smoothly, but, instead, toward saving our planet's environment. So I looked for ways to change careers, and decided to go to law school and become an environmental lawyer.

I graduated from Loyola Law School in Los Angeles in 2010, took the California Bar exam, and started a non-profit law firm called Advocates for the Environment. My practice, described in more detail in the Law chapter of this book, primarily uses the California Environmental Quality Act to force development projects to be net-zero in terms of their emissions of greenhouse gases. This is a large, concrete and quantifiable benefit towards fighting the climate emergency, and results in much more impact than I could possibly have through changes in my personal life.

At some point during the last decade, after I'd started the original, non-climate-oriented version of this book, I realized that climate is by far the most important challenge of our age, and that few of the books and articles dealing with it frame it the way it should be framed, as primarily an ethical issue. So my purpose in writing this book is to present a fairly complete survey of the many aspects of the climate problem through a primarily ethical lens.

Useful information when reading this book

The overall scheme

The main thesis of this book is that ethical and moral considerations compel us to act decisively, and as quickly as possible, to avert the effects of climate change. I hope I will have convinced you of that by the end. But the book has another purpose as well, which is to provide a fairly comprehensive, high-level summary of bodies of knowledge that contribute to an understanding of the problem and its potential solutions. The chapters all reference sources that you can consult for more in-depth information.

Chapter 1 is about Ethics, and how our traditional morality needs to be updated to deal with issues forcefully presented by the climate crisis. The most important of these issues are how we value non-human animals and future generations.

Chapter 2 is a summary of the current science on climate change.

Chapter 3 is about sustainability, the idea that we shouldn't be depleting Earth's resources at such a high rate that they will leave future generations environmentally impoverished.

Chapter 4 is about the economics of climate change. Many environmentalists still incorrectly believe that switching from fossil fuels to renewables will not cost anything because renewables are cheaper now. I present a simple economic model, and show that phasing out the burning of fossil fuels by 2050 is the economically optimal course of action. I also provide estimates of what this is likely to cost.

Chapter 5 is about climate law. It covers litigation, legislation, and treaties in the US and internationally. It's the broadest summary of this topic that I'm aware of.

Chapter 6 is about climate politics, but also discusses some other issues, like the relationship between climate and environmental justice.

Chapter 7 is the conclusion, which sums up the steps we need to take to deal with global heating. We need a stronger international cooperation, and a stronger climate treaty, to make sure that the major emitters reduce their emissions before the planet heats up too much.

International viewpoint

This book tries to take an international, global perspective, because the climate emergency is a global problem; no one country can solve it alone. In fact, it will be very difficult to deal with unless we have the strong cooperation of the big emitters: China, India, the US, and the EU. I'm an American, and an American lawyer, so I know the US better than other countries, and tend to give the US a bit more prominence than it deserves in these discussions. I've had a life-long interest in Europe, especially France, and I've travelled extensively in Europe and have studied EU and international law. I speak some European languages. I've travelled to China and Latin America, but am pretty ignorant about India, other parts of Asia, and Africa.

The term "state" can be confusing in the context of this book. Our US federated government is made up of 50 states, plus a federal government. Under the US Constitution, the 50 states retain some sovereignty, since our government started out, like the EU, as a federation of sovereigns. But the international community uses the term "State" to refer to a country, like France. I'll try to distinguish the two terms by capitalizing the international one, and will try to supply a context that makes it clear which one I'm talking about.

The Anthropocene

Geological scientists reviewing the fossil record find strata, and, based on evidence in these strata in the rocks, divide the geologic past into epochs and ages. The Jurassic epoch, for example, lasted from 201.4 to 145 million years ago. Recently, geologists have suggested a new epoch, called the Anthropocene epoch, signifying humans' newly found power to act in ways that materially change this planet, and which leave impressions in the fossil record. Various dates have been proposed for the start of the Anthropocene, such as the World War II atomic bombings, but I prefer the industrial revolution as the start date.

The point in making this difficult word a part of the title of this book is that it signifies a major turning point in human and Earth history. Before the industrial revolution, we humans didn't need to worry nearly as much about permanently harming the planet because humans didn't have the power to do that. Now we do, and we therefore have a responsibility to manage our planet and make sure our actions don't permanently degrade it.

The metric system

I use the metric system in this book, partly because of the book's international outlook, and partly because the metric system is the system that scientists use. If you're unfamiliar with it, it's worth getting to know. There are several handy online services that do conversions between metric and imperial units. The following is a short glossary of metric units, with translations into their imperial equivalents.

- Length: the meter, about 39 inches, is the basic unit, with a centimeter being a hundredth of a meter (roughly a quarter of an inch), and a millimeter being a thousandth of a meter. A kilometer is a thousand meters, about 0.6 miles.
- Area: square meters are obvious; a hectare is 10,000 square meters, about 2.5 acres.
- Volume: cubic meters are obvious, as are cubic centimeters. One cc is about 1/30 of an imperial fluid ounce.

- Mass: a kilogram, a thousand grams, is about 2.2 imperial pounds. A gram is about 1/28 of an ounce. A metric tonne (the standard unit used to measure greenhouse gases) is a thousand kilograms, about 1.1 imperial tons. A gigaton (Gt) is a million (metric) tons, equal to a petagram (Pg).
- Temperature: a degree Celsius or Kelvin (1°C) is 1.8 degrees Fahrenheit, so a warming of 3°C would be a warming of 5.4°F. The difference between Celsius and Kelvin is the zero-point. For Celsius, it's the freezing point of water (0°C is 32°F); for Kelvin, it's absolute zero, -273.15°C.
- Power: the common unit, even in countries like the US and UK that generally use the imperial system, is the watt, which is a metric unit. This is a unit of energy generated per unit time. A kilowatt is a thousand watts.
- Energy: the basic metric unit is the joule; the most common unit is the kilowatt-hour, which is actually a metric unit because the watt is a metric unit. A kilowatt-hour (kWh) is, as you might suspect, the amount of energy used in an hour by an appliance that draws a kilowatt of power. A kilowatt-hour is equivalent to about 3.6 million joules. A British Thermal Unit (BTU) is another unit of energy, equivalent to 1,055 joules or 1/3412th of a kilowatt-hour. A quad is 10^{15} BTU.

Scientific notation: Scientists and engineers use powers of ten to represent very large or small numbers. "10^3" means ten to the third power, which means ten multiplied by itself three times, or 1,000. 10^6 is a million. A typical number in scientific notation is 3.14×10^4, which is 31,400.

Acronyms

There are a lot of acronyms in the climate field. Here is a list of most of the acronyms, initialisms, and abbreviations used in this book.

- ACLU – the American Civil Liberties Union
- AFOLU – Agriculture, Forestry and Other Land Use
- AGI – Adjusted Gross Income
- AI – Artificial Intelligence

- BCE – Before Common Era, or B.C.
- CAA – US Clean Air Act
- CBD – Convention on Biological Diversity
- CCS – Carbon Capture and Storage
- CE – Common Era, or A.D.
- CEDAW – Convention on the Elimination of Discrimination Against Women
- CEQA – California Environmental Quality Act
- CFCs - Chlorofluorocarbons
- CMIP – Coupled Model Intercomparison Project
- CO_2 – Carbon dioxide
- COP – Conference of the Parties
- COP27 – the 27th Conference of the Parties of the UNFCCC
- CTBT – Comprehensive Nuclear Test Ban Treaty
- CWA – US Clean Water Act
- EIA – Environmental Impact Assessment
- EIR – Environmental Impact Report
- ENSO – El Niño-Southern Oscillation
- EU – European Union
- EV – Electric Vehicle
- G20 – Group of 20 most-developed countries
- GDP – Gross Domestic Product
- GHG – Greenhouse gas
- GWP – Global Warming Potential
- HFC - Hydrofluorocarbons
- ICJ – International Court of Justice
- IPCC – Intergovernmental Panel on Climate Change
- IRA – Inflation Reduction Act
- LEED – Leadership in Energy and Environmental Design, a green-building certification program
- MND – Mitigated Negative Declaration
- MTCO2e – Metric tons of CO2 equivalent
- NDC – Nationally determined contribution under the Paris Agreement
- NEPA – US National Environmental Policy Act
- NO_2 – Nitrous oxide

- NPV – Net present value
- OECD – Organization for Economic Co-operation and Development, an international organization of 51 relatively developed countries
- P.p.m. – Parts per million
- PFC - Perfluorochemicals
- R&D – Research and Development
- SCC – Social Cost of Carbon
- SDGs – UN Sustainable Development Goals
- SER – Sustainability-Efficiency-Renewables framework
- SSPs - Shared Socioeconomic Pathways
- TEU – Treaty on European Union
- TFEU – Treaty on the Functioning of the European Union
- UNCLOS – United Nations Convention on the Law of the Sea
- UNEP – United Nations Environmental Program
- UNFCCC – United Nations Framework Convention on Climate Change
- USC – U.S. Code of laws
- VCLT – Vienna Convention on the Law of Treaties
- VMT – Vehicle Miles Travelled
- WGI, WGII, WGIII – the three working groups of the IPCC synthesis-report process

Chapter 1
Climate Ethics

"Ma philosophie est en action, en usage naturel et present, peu en fantaisie." — Montaigne

I'm an Earthling: a denizen, partisan, and patriot of the Planet Earth. My main allegiance is to Earth. I live in the neighborhood called Shadow Hills, in the City of Los Angeles, State of California, in the country named the United States of America, located in North America, but my allegiance is to the Earth as a planetary whole, not to these sub-places or sub-groups. The principal reason my Earthling patriotism isn't more widely shared is that Earth includes everyone and everything; most people need an enemy, an out-group or opposing team to cement their loyalty to their in-group. It would be easier to be patriotic for the Earth if there were another team that we could be patriotic against. Extra-terrestrial space invaders landing their flying saucers on the Capital Mall would make us all Earth patriots.

We need a new ethics for a new phase of human life. We have recently acquired the ability to significantly harm the planet Earth, or at least life on Earth. Earth is a ball of rock spinning through space with a thin surface film containing life. Nothing we are doing will hurt the ball of rock itself, but human activities, for the first time, are significantly harming at least the higher orders of life, mostly through climate change. Unless we reverse course quickly, these harms will increase and become permanent. How we deal with this crisis will be the most important collective decision made so far in human history.

This book is primarily about global warming, an emergent crisis. Deciding how to deal with this crisis implicates a number of moral and ethical issues, and requires new perspectives on some of them. The purpose of this chapter is to outline the ethical framework I apply to the problem. I don't expect that you, the reader, will adopt this framework wholesale. But this brief tour d'horizon may prompt you to consider and develop your own views on these important moral questions.

Life on Earth

What is a day of life worth? I've lived about 25,000 days so far, and they're flying by fast. I have just a few thousand days left, but haven't internalized the fact that they'll end. If I knew I had only ten days of life left, each of those days would be precious. A platitude urges us to enjoy each day as if it is our last. This makes no sense; I would be completely selfish about how I used my time if I knew I were going to die tonight at midnight. But that is not how I want to live my life. I have important things to do that take a lot longer than a day. Such as finishing this book. There's work to be done, helping out my fellow humans and the rest of life on Earth.

Surely life itself is a source of moral values, an ethical *cogito ergo sum* — I live therefore I value life. My enthusiasm for life begets fellow feeling for other living creatures, not just humans. Their lives have value to them, just as mine does to me. This concept is a lot like Albert Schweitzer's Reverence for Life, the basis for his ethical system: "Ethics is nothing other than Reverence for Life. Reverence for Life affords me my fundamental principle of morality, namely, that good consists in maintaining, assisting, and enhancing life, and to destroy, to harm or to hinder life is evil."[1]

Valuing our own lives gives us sympathy for the lives of others, which also must be valued. The point of Buddhist mindfulness, at least for me, is that we shouldn't take our lives for granted. We should pay attention and make sure we experience what's happening to us right now. Later in life, especially, the years, and even the decades, can click by unnoticed. We get caught up in our daily problems and routines and don't pay attention to life itself, and what a wonderful gift it is. The joy in life is something we can share with our fellow humans and other living creatures on Earth.

As far as we know, life exists only on our planet Earth, and is inextricably linked to Earth. From a distant point of view, life is a scum on the surface of this ball of rock. If Earth were scaled down to basketball size, the atmosphere, which is about 50 km deep, would occupy the space of just a

---

[1] Albert Schweitzer, *Civilization and Ethics*. 3rd ed. (London:Adam & Charles Black, 1961).

millimeter—less than a sixteenth of an inch—above the surface of the basketball, and everything we know, up to Mt. Everest, would fit in the lower tenth of this millimeter, less than a hundredth of an inch.

We've seen the pictures from space confirming our relative insignificance— Earth as a mote circling the speck that is our sun, off in a remote corner of our Milky Way galaxy, which is comprised of 200 billion other suns, and is just one of a hundred billion galaxies in the universe.

But, without life, what does it all matter? Stars revolving around stars in galaxies revolving around galaxies are just a huge, empty machine, insignificant without life. Life on Earth matters, more than the statistics suggest.

Will we ever find other life? A major theme of modern physics is the limited information available to us. Quantum mechanics tells us we can never exactly measure, predict, or simulate the smallest systems of atoms and elementary particles. Thermodynamics teaches that entropy—disorder, the opposite of information—inexorably increases. Special relativity precludes anything that could convey information from travelling faster than the speed of light, putting huge regions of space-time beyond our reach. Even if there is other life somewhere we may never find it.

The likelihood of life elsewhere is the product of the teeny chance of life arising on a planet and the potentially huge number of suitable planets. We don't know either of these numbers. We surmise that once life gets started, by a chance concatenation of self-reproduction and metabolism, the process of natural selection will allow it to develop and diversify. But how likely is life to start, even in conditions identical to those on Earth? And, once life has started, how likely is it to develop into an intelligent civilization? Maybe the occurrence on Earth is a freakish coincidence, unique in the universe. Or maybe it's fairly common, even inevitable. We don't even know whether life must use our water-based chemistry.

The question of why astronomers have seen no direct evidence of life on other planets, when there are almost certainly huge numbers of planets that could support life is known as the Fermi Paradox. The physicist

Enrico Fermi asked this question of other eminent physicists at lunch in 1950.[2]

As far as we know, we're the only game in town. The Earth is the only place we know of where life has arisen. That makes us and what we do important, even on a cosmic scale.

The big picture

I'm a long-term and big-picture type of guy; I always want to step back to consider the larger context. Climate change is a big-picture issue, so the issue fits with my natural predilection. The big picture for climate change is that it will affect every person, and most plants and animals, on the planet, and will have impacts, even after we get it under control, that will last for hundreds or thousands of years. These far-reaching effects mean that the impacts we can feel across the globe today—increased droughts, wildfires, hurricanes, crop failures, heat waves, sea-level rise—are only a small fraction of the eventual impacts.

The big scale of the problem makes it hard for humans to understand. We're used to thinking in small numbers, up to thousands, but when we get to millions, billions, and trillions, we have no personal, physical experience with such large numbers of things, so our minds tend to lump them together as "a big number." This psychology is illustrated by the largest number for which the ancient Greeks had a specific name, the "myriad," which means 10,000. Once a number got to be over a thousand, the Greeks would call it a myriad, to indicate it was beyond human comprehension. But we can't do that—timescales and physical scales based on large numbers are important for understanding climate change science.

I try to have a global, rather than US-centered perspective, and to consider the impacts of what we're doing now from a vantage point in the year 3000. Seeing the climate-change big picture means taking a global viewpoint that considers everyone on the planet, factoring the interests of non-human animals into ethical decisions, and taking into account the interests of

---

[2] "Fermi Paradox," in *Wikipedia*, https://en.wikipedia.org/wiki/Fermi_paradox.

future generations. But, when doing so, I try to keep in mind the impacts on individuals. Global heating is harming individual humans and animals in different ways, depending on personal characteristics and geographic location. And the harms are not distributed equitably. We cannot afford to be callous or dismissive about these individual impacts when we adjust our mental magnification to take in the big picture.

An existential decision

How do we decide what our values are? I've just sketched out a basis for an ethics based on valuing life on Earth. We all come from different backgrounds and have different bases for our ethical decisions. Yet, as a society, we mostly agree on a common ethical framework.

Adherents of monotheistic religions, such as Christians, believe their ethics come from God. The source of ethics for Buddhists is natural, not supernatural: the teachings of Buddha and the accumulated wisdom of subsequent practitioners of Buddhism. Secular humanists believe that humans can develop an effective code of ethics through rational means and empathy.

In the end, each of us must choose his or her own ethics, either as part of a decision to adopt a religion, or apart from religion. Many people choose a religion not on the basis that they think it is actually true, but because they like its ethical and spiritual ambiance. Or they're born into it. If I believed, as Christians say they believe, that my actions during my lifetime would determine whether I was consigned to heaven or hell for all eternity thereafter, I would devote every speck of my time on earth to getting into heaven. Most Christians do not do this, so I deduce from their behavior that they do not really believe in heaven or hell. To find out what they really think, look at what they do, not what they say.

The basic idea of existentialism is that existence precedes essence for humans. About 2,500 years ago Plato came up with the idea of forms, which are the universal patterns or templates used for instantiating particular objects. The form for a table is the essence of a table, and defines "tableness," so we can use it to decide whether a particular object is a table.

A craftsman has the table-form in mind when creating a table. The form—the essence—precedes the coming into existence of the instance of a table, so, for most things, essence precedes existence.

Is there such a form or essence for humans, which defines human nature? According to the Book of Genesis, God created humans in the image of God. This implies a form for mankind, namely God himself or herself, though obviously the humans God created in his or her image are not identical to God. Jean-Paul Sartre, who came up with the idea that existence precedes essence for human beings, and thus created existentialism, says, essentially, "no, there is no human nature but what we each create for ourself by taking responsibility for our lives and deciding what our own nature should be." In a sense, we create ourselves through our moral choices.

This view does not take the science of evolution into account. Humans are adapted to be hunter-gatherers, running around in small groups finding food in the wild. We were like that for hundreds of thousands of years until we invented agriculture about 12,500 years ago. Agriculture allowed us to develop larger communities of people, which, in turn, led to the invention of writing, which allowed us to preserve culture and knowledge from one generation to the next. We have probably continued to evolve and adapt following our invention of agriculture, but most of our human nature was formed in the much-longer hunter-gatherer period. So when we are making existential choices as to who we want to be, we are not starting with a blank slate; we have an in-built nature that colors and limits our choices. We often don't perceive this human nature because it's there all the time in everyone. A major theme of this book is that the human nature that evolved during our hunter-gatherer existence is not well suited for our current situation.

But the fact that there is a human nature that influences and limits us does not mean that we don't make existential choices. And one of the most important choices we each must make is what our values are. What is important to me? What should I live my life for? What ethical rules should I follow? These are existential choices, in which each of us has the opportunity to define himself or herself. The existentialists would add: we're responsible for our choices, and, through those choices, we're responsible for the world. It's a lot of responsibility.

Most of us don't expressly make these big choices, at least not all at once. If you grow up Muslim, you are taught a system of values and ethics, which you adopt, at least initially, by default. A big event in your life might cause you to re-think part of your belief system and make a change. People generally adopt a set of values by default, and then those values evolve somewhat based on experience.

In this section of this book I am advocating a moral system that intrinsically values Life on Earth, and gives particular value to sentient life. We can all choose to do this, and these values are not incompatible with most other moral systems. For example, monotheists can believe that God wants us humans to be good stewards of creation, i.e. of Life on Earth. Buddhists are already of the mindset that life is fungible through reincarnation—we could come back as a dog or a mouse next time, so there is no categorical distinction between humans and other animal species. Thus all life is valued.

Philosophy must be the framework

A friend of mine suggested we just need to do a cost-benefit analysis to figure out what to do about the climate crisis. Economists are forever trying to cast everything in economic terms, including the ethics of climate change. The advantage is that, once all important considerations are expressed in economic terms, we have a suite of economic tools to analyze the trade-offs. And there is an underlying assumption that most of what's valuable in life can be expressed as money, and even that the main function of a human is to make money and improve one's economic welfare.

I would counter that the material side of human life is not the most important. Residents of some countries with far less income than Americans rate higher on polls' happiness surveys. Friendships and other human relationships, time in unsullied natural settings, freedom from insecurities, avoiding increased threats from drought, storms and other extreme weather events, wildfires, and sea-level rise, can be just as important, or more important than economic welfare.

The framework for analyzing climate change must be philosophical—ethical and moral. Economics is an important part of the analysis, but not the main part. Similarly, science and the law are important parts of the picture, but neither of them can provide the top-level frame. Science is our way of analyzing facts in the physical world, and is important for understanding the mechanisms and impacts of the climate crisis. But it sheds no light on ethical issues. Law is our high-level rulebook, but it's based on our ethical and moral choices about how we should get on together. We formulate laws after we've come to a consensus or compromise on the underlying ethical issues. This book contains chapters on economics, science of climate change, and the law, but this is the most important chapter, providing an overall framework based on ethical and moral considerations.

No taboos

I will venture into areas that may make you, the reader, uncomfortable or angry. People get emotional about their values, especially when those values are challenged. I am trying to have a free discussion, for all types of people: religious, and non-religious, Western and Eastern, rich and poor, educated and not. You all come to this discussion with different values, but that shouldn't prevent us discussing them. You will almost certainly disagree with some of what I say, but try to keep an open mind, and don't consider some areas of moral thought taboo. I'll try to do the same.

Recently, while driving with my wife, Benita, we stopped at a traffic light on the main street near our house. This particular light was connected to a radar unit that sensed how fast we were driving. We were a bit over the speed limit and so the traffic light changed to red, in an attempt to incentivize us to stay below the speed limit. "Forcing me to stop and then start up again wastes gas—it's bad for global warming," I said. Benita replied "But it saves lives by motivating drivers to drive slower." "How do you trade off a human life against global warming?" She was shocked I'd even ask the question.

She, like most people, values humans first. Everything else is a distant second. There is an assumption in her reply that a human death trumps everything, including environmental concerns. We're not supposed to balance human lives against other things we value such as money. My comment suggested that perhaps the environment ought to come first in some situations, even before human life, or at least that the effects of a death on the environment ought to be factored into our safety policies. I haven't changed her mind, but I think I've made her realize the issue of whether the environment can trump human lives in some situations should be open for thought and discussion.

Many of us were taught as children to trust our consciences, and rely on internal moral compasses, our feeling for what's right and wrong, when making moral decisions. That advice will not stand us in good stead in this new age, where the higher orders of Life on Earth, which we humans need for our own survival, are being threatened.

The moral compass is a fairly reliable indication of the time because we have absorbed values from our education and culture, so our emotional reaction when there's a question of values reflects those general societal values. If we're rethinking those values, as I suggest we do in this book, the result may be a recalibration of the compass.

We must adjust our ethics and adopt new values to fit our new Anthropocene age, which started when we acquired the power to significantly harm the higher orders of life on our home planet Earth. We're well down the path of destruction and need to find a new way. New societal norms will be based on our moral values. The individuals in our society will never have identical values, but we need to develop a consensus large enough to allow us to forge a new path.

Animals in the moral universe

The tree of life has millions of species as its leaves. Many of them are single-cell species such as bacteria, and there are plenty of other non-sentient species such as plants. We really don't know for sure, to within an order of magnitude, how many species there are, or the relative proportions of plants, animals, insects, fungi, bacteria, etc.

For most of us, there is no moral harm in hurting or killing non-sentient life, though those who follow the Jain religion take care to avoid undue harm to plants and even to microorganisms when preparing food. Non-sentient life doesn't endure pain or suffering when damaged or killed. At the other extreme, anyone who has witnessed a wounded dog or bear will agree that sentient beings, at least higher-order ones, can suffer. They have other feelings as well.

Nerves make animals sentient, so nerves are the admission ticket to the moral universe. Nervous-system sizes are approximately:

- a few hundred neurons for a worm
- about a thousand, for a jellyfish
- hundreds of thousands, for insects
- hundreds of millions, for mammals
- billions, for primates
- almost a hundred billion, for humans

Charles Darwin noticed that a worm could modify its behavior in withdrawing from sudden illumination depending on whether its attention was occupied with some other matter. For him, this showed "the presence of a mind of some kind."[3]

The higher an animal is on this scale, the greater the degree of sentience, but it's a crude yardstick. African elephants have about three times as many neurons as humans, and long-finned pilot whales have twice the number of cerebral-cortex neurons. We assume these animals with more neurons don't think as well as humans, and don't have the sense of self-consciousness that we do. But maybe they're smarter than we are—they look at humans and see hyper-active busy bees, caught up in building, organizing, scurrying around, always busy, and think "what's it all for? I can have a happier life if I just keep my mouth shut and carry on with my nice quiet life in the jungle or sea."

It's difficult for us to understand the subjective states of non-human animals, and more difficult for animals lower on the neuron scale. We understand other humans' feelings only by analogy to our own. We

---

[3] Charles Darwin, *The Formation of Vegetable Mould through the Action of Worms: With Observations on Their Habits* (London: John Murray, 1881).

presume that when their behavior is similar to ours their feelings are similar, too, but we have no way to verify this. My best friend might be living in a completely different subjective universe than I live in, though our outward behaviors are similar.

The problem is greatly increased for other species. We might have a rough idea of the pain a wounded dog is feeling, but what is it like for a wounded ant? Does an ant whose lower body has been squished feel the same agony as a human whose legs have been crushed? The ant could be writhing as a purely mechanical reaction to injury, not from any subjective feeling of pain. The philosopher Thomas Nagel wrote an influential paper in 1979 on "What is it like to be a bat?" He used the bat as an example because its senses, such as sonar, are so different from ours, for whom optical vision is most important. The famous French philosopher René Descartes thought that animals were automata because they possessed no souls and were thus incapable of feeling pain.

We all accept that pain and suffering for humans is a bad thing, and I submit that pain and suffering for animals is also bad, and that such pain and suffering exists in proportion to the animal's relative position on the neuron scale. A bacterium doesn't suffer when it is damaged; a dog does.

We could try to distinguish animals from humans on the basis that the latter are conscious, and the former are not. By "conscious," we usually mean "having an awareness of one's own existence, sensations and thoughts" or "capable of thought, will or perception." We're all subjectively aware of our own consciousness. We feel things and have a train of verbal pitter-patter running in our brains most of the time. We presume, by analogy, that other humans have the same mental world we do, but that analogy weakens when applied to animals. There is, however, a lot of evidence that animals have some form of consciousness. We shouldn't make a hard distinction between humans and other animals on the basis of consciousness.

We could also distinguish on the basis of emotions. We know from our subjective experience that humans possess emotions. It's fairly obvious to dog owners that dogs have emotions, too. How far down the tree of life does this go? How complex does a brain have to be to have emotions? Mammals have emotions, and invertebrates probably have them, too. Some researchers have even suggested that honeybees have emotions.[4] Again, this test results in a graduated spectrum, though it's harder to be sure that emotionality diminishes for smaller brains with fewer neurons. It's an adaptive characteristic, like cognition, but may not require the complex brain needed for cognition.

Traditional humanism also puts humans first, as its name implies. "Man is the measure of all things," said Protagoras, a Greek sophist who lived in the fifth century BCE. Secular humanists take human concerns instead of God as the source of their ethics. Humanism doesn't focus on whether humans are the only life forms that deserve moral consideration, but their approach implies they are. This approach was taken in Western philosophy: Aristotle, another ancient Greek philosopher, argued that animals lack reason. Descartes' dualistic solution to the mind-body problem was that the physical world was mechanistic, but mind was not part of or based on the physical world; it was closely connected with the soul, a link to God that only humans have.

In traditional Christian theology, humans have souls and animals do not. This dichotomy removes non-human animals from the moral universe. According to the Book of Genesis, God put plants and animals on earth for humans to exploit; he gave man dominion over Earth and everything on it. Psalm 115 tells us "the heavens are the Lord's, but the earth hath he given to the children of men."

Some Eastern religions have taken the contrary approach. Buddhism teaches that humans can be reincarnated as animals and vice versa. There is no categorical distinction between them. The Hindu concept of *ahimsa*, which mean "non-injury" or "non-killing," was adopted into Jainism and Buddhism.[5] It prohibits violence by physical act, or by words or thoughts., and it applies to violence against all forms of life, including animals, plants,

---

[4] M. Bateson et al., "Agitated Honeybees Exhibit Pessimistic Cognitive Biases," *Current Biology* 21, no. 12 (2011): 1070–73.

[5] "Ahimsa," in *Wikipedia*, https://en.wikipedia.org/wiki/Ahimsa.

and microbes. Mohandas Gandhi, an anti-colonial Indian nationalist, adopted ahimsa as the basis for his non-violent civil disobedience. The Jains have a moral scale somewhat similar to the neuron-based one I've proposed in this book: animals are ranked by number of senses, from one-sensed ones that perceive only touch, up to five-sensed animals like humans.

The approach taken by our Western society and legal system is in line with the Western, Christian view: Humans have moral standing, and non-human animals do not. Legally, pets such as dogs and cats, are property, like cars, and like human slaves, before slavery was outlawed. Similarly, wild animals have no legal rights; humans are in a completely different moral category from non-human animals.

The polar opposite view is American environmentalist Aldo Leopold's land ethic, which advocates moral rights of "soils, waters, plants, and animals, or collectively: the land."[6] In my view, this goes too far. Do the soils and waters of the planet Venus have moral rights, even though there is no life on the planet? How about asteroids?

In this book, life on Earth includes all life on the planet. Species that have neurons participate in the moral universe in proportion to their sentience. We have no moral duties toward species without neurons, but those non-sentient lifeforms are valuable as part of the planet's ecological fabric. The sentient creatures cannot live without them. We have responsibility for managing them as we manage rivers and land use, and, indeed, the Earth itself.

Humans or Earth – which is more important?

Is Earth here just to be exploited as a resource for humans, or does it have greater importance than that? Environmentalists like to say that humans are one species out of millions, just another thread in the fabric of life on the planet. This may have been true when we were hunter-gatherers living

---

[6] Aldo Leopold, *Sand Country Almanac, Special Commemorative Edition* (Oxford: Oxford University Press, 1987), 204.

in our original ecological niche. But now, through technology, we have the ability to dramatically change the Earth's ecology. Since humans uniquely possess this power, humans are special and unique, even if we have no greater moral standing than other animals.

The anthropocentric (human-centered), instrumental view is common today: that the natural world is just a resource for humans to exploit, and it's the humans that matter, not the rest of nature. Our terminology implies this view. "Environment" means "the objects or the region surrounding anything,"[7] because "environ," from Old French means "around." What is being surrounded, or "environed"? It's us, of course, humans. The environment is the matrix in which we live, defined in relation to humans.

On a practical level, the dichotomy between humans and the Earth is false, because we are dependent on the Earth for our existence. Some prominent scientists such as Stephen Hawking say we should have a backup plan, in case Earth becomes uninhabitable.[8] This type of thinking started back in the 1970s when there was a movement to build colonies in space, man-made cylinders that would provide 100 square miles of living space, each allowing 10,000 people to live in comfort.[9] During that era, Biosphere 2 and BIOS-3 were built, in Arizona and Russia respectively, to experiment with closed biological systems that could sustain human life. Those experiments largely failed, showing that it is not easy to pick a subset of plants and animals to form a small artificial closed ecosystem capable of supporting human life.

The Biosphere failure should serve as a warning against our willy-nilly destruction of species and ecosystems through climate change, toxins, and habitat degradation. We humans are part of the global ecosystem. We blithely assume it's infinitely resilient, even though we don't understand how it works. We can't live without it. We destroy it at our peril.

---

[7] "Environment," in *Oxford English Dictionary*, 1971.

[8] Naomi Klein, *This Changes Everything: Capitalism vs the Climate* (New York: Simon & Schuster, 2014), 288.

[9] Stewart Brand, *Space Colonies* (Middlesex: Penguin Books, 1977).

Until we can design a small, closed ecosystem supporting human life, and solve a host of other technological and economic problems, we have no backup or Plan B to replace Earth. We have no place to go if we render our planet uninhabitable. The Earth is as important as human life because, at least for the foreseeable future, human life is dependent on the Earth.

Some environmentalists treat the entire earth as a living organism with the ultimate moral standing. For example, the Gaia hypothesis holds that Earth is a self-regulating complex system which seeks an environment optimal for contemporary life.[10] According to this theory, organisms co-evolve with their environment. They influence their abiotic environment and that environment in turn influences the biota by Darwinian processes. The hypothesis is scientifically questionable,[11] but the underlying philosophy fits in with other strands of environmental philosophy that give moral standing to the Earth.

Deep Ecology, a philosophy invented by Norwegian philosopher Arne Naess in the 1970s, advocates that the living environment as a whole should be respected and regarded as having inalienable rights to flourish. In Deep Ecology, "the right of all forms [of life] to live is a universal right which cannot be quantified. No single species of living being has more of this particular right to live and unfold than any other species."[12] This is the polar opposite of anthropocentrism: humans have no more moral standing than any other species, including fungi and insects. As just one among the tens of millions of species on Earth, the Earth is vastly more important than humans.

This, too, goes too far. A great deal of the moral significance of Life on Earth comes from humans, and most of the rest of the Earth's moral significance comes from sentient life. 800 million years ago, when the only life on Earth

---

[10] James Lovelock, *The Vanishing Face of Gaia: A Final Warning* (Basic Books, 2009), 255.

[11] Toby Tyrrell, *On Gaia: A Critical Investigation of the Relationship between Life and Earth* (Princeton, NJ: Princeton University Press, 2013).

[12] Arne Naess, "The shallow and the deep, long-range ecology movement. A summary," *Inquiry*, 16, no. 1–4 (August 29, 2008): 95–100.

was single-cell life such as bacteria, how much moral significance did it have? I'd say its significance came from its potential to eventually develop intelligent life.

Our answer concerning the relative importance of humans and Earth's ecosystems as a whole depends on whether there is life, especially intelligent life, elsewhere. If we knew there were millions of other planets with intelligent life, the importance of both humans and Earth in the overall scheme of things would diminish considerably. Humans would then be our home team, important to us in the way a high school's basketball team is more important to the school's students than other teams. Trashing earth so as to render it uninhabitable by humans would not be of universal significance.

Another extreme view is that humans are a plague on the Earth. It's true that, when we invented agriculture around ten thousand years ago, we escaped our ecological niche as top-predator hunter-gatherers. That niche limited our numbers to a few million because of the scarcity of food, but agriculture allowed us to break through our ecological barrier and increase our population far beyond its natural limit. The human population is now over eight billion and increasing. Humans are degrading the environment by using natural resources at a rate much higher than they can be replenished. If we just cared about natural ecosystems, the planet would be better off without humans. Some environmentalists take this position and, by doing so, are showing they value the Earth much more than human life. For the reasons given above, I disagree. Humans are an important part — probably the most important part — of life on Earth.

Another skein of environmentalism wants us to go back to living in harmony with nature, as we did when we were hunter-gatherers. For example, the American nature writer Barry Lopez counsels us to tap into the wisdom of pre-Columbian native Americans and wild animals to develop a new, respectful relationship with the landscapes we live in.[13] He rues Hernando Cortés' 1520 destruction of the Aztec capital, Tenochtitlán,

---

[13] Barry Lopez, "The Passing Wisdom of Birds," in *Crossing Open Ground* (New York: Vintage, 1989).

a large city he exalts as an example of such harmony. He claims five hundred Native American cultures lived in enlightened intimacy with the land, each in a different way. And he counsels us to observe animals in the wild, to absorb their symbiosis with the landscape.

I backpack in the wilderness to animate my inner hunter-gatherer. Roaming the landscape is what evolution adapted us to do. It's our nature. We did it for hundreds of thousands of years before we settled down to a rooted agricultural existence. There are those who think our lives were better before the agricultural revolution.[14] But I love civilization and the arts too much to want to go back. And it's unrealistic to think we can return to that kind of way of life. It would require too much economic sacrifice and a huge reduction in population.

We must instead find a middle way, a way to maintain a sufficient economic standard of living that does not degrade the Earth. This middle way will use less resources and will therefore, by economists' measures, result in a lower standard of living. But it can also result in a richer life through a rebalancing of values toward friends, community, and leisure.

Earth is our parent; Earth gave birth to us. We behave toward Earth as children behave toward their parents. Parents are so much more powerful than young children and control their world. They behave in ways that children cannot fathom, so they seem like forces of nature, part of the cosmos. It doesn't occur to young children that they could hurt their parents; they don't identify with them as people, so they don't take into account the effects of their actions on their parents. A big part of the process of becoming an adult is coming to grips with the reality that one's parents are people like us and have similar motivations and fallibilities. We humans, up until recently, have treated Earth, our parent, as if we were young children. We haven't realized that Earth has needs and that our actions can hurt our parent. We need to grow up.

---

[14] Yuval Noah Harari, *Sapiens: A Brief History of Humankind* (Signal Books, 2014), 77–81.

Human nature

What is our human nature? The following are the most important developments that have contributed to our human nature:

1. Genus *homo* evolves from apes – about 2 million years ago. *Homo erectus* was a hunter-gatherer.
2. Homo Sapiens evolves as a distinct species of primate – about 200,000 years ago
3. Humans start developing language and complex reasoning – about 100,000 years ago
4. Humans develop agriculture – about 12,500 years ago
5. Humans develop writing – about 5000 years ago
6. The industrial revolution – the start of the Anthropocene era – about 1800 CE
7. The computer revolution – about 1950 CE

Some biologists argue that intelligent life would evolve on another planet in a very similar way to its evolution on Earth.[15] This is because of evolutionary "convergence," which often generates the same solution to a problem on several evolutionary pathways. It's fairly evident that intelligent extra-terrestrials would share many human characteristics. The differences would be illuminating, though. Our sensors are not the only ones possible—bat-like sonar, "eyes" that see a different portion of the electromagnetic spectrum, and an organ that acutely senses ground vibrations are all possibilities. Other life would presumably be based on some DNA-like self-replicating mechanism. Would the chemistry necessarily be based in water and carbon like ours? We don't know.

It would be in the area of the mind that the differences would be most interesting. If we could have congress with a dozen races of extraterrestrials, we could much better answer the question of what is human nature; we'd be able to see which characteristics we share— presumably the convergent characteristics necessary for development of

---

[15] Simon Conway Morris, *The Runes of Evolution: How the Universe Became Self-Aware* (West Conshohocken, PA: Templeton Press, 2015).

intelligence—and where we differ. We'd be able to consider which of the different characteristics are superior for building the kind of society we want to build.

The field of evolutionary psychology deals with how our minds were formed by adapting to our environment.[16] Humans have "stone-age minds" because our ancestors spent millions of years as hunter-gatherers before we developed agriculture a mere 12,500 years ago. Our brains are designed to hunt and forage for food, and to live in groups of a few dozen individuals with more cohesive cultures.

This is our human nature, which is unsuited in many ways to dealing with the problems we face now. First and foremost, we're wired to focus on the short term: getting food now, fighting off the attack of an enemy or a wild animal. Second, we're designed to work in relatively small groups—a few dozen people—and don't naturally extend our trust to a wider circle. Third, hunter-gatherers were mostly nomadic, moving frequently, living in temporarily constructed shelters, with few permanent possessions. They fit into their ecosystems the same way as other predators, and their numbers were limited by the same population dynamics that controlled other top predator species.

To better live under the circumstances of modern life, human nature should be changed to be (1) more focused on the long term; (2) less tribal and nationalistic, and more open to working and collaborating with larger groups and different types of people; and (3) less competitive and more cooperative.

It's unclear how much and how fast human evolution will continue. This question is controversial among scientists.[17] We're mixing everyone much more broadly across the globe than we have ever done, and we're

---

[16] Leda Cosmides and John Tooby, "Evolutionary Psychology: A Primer" (UCSB Center for Evolutionary Psychology, 1997), https://www.cep.ucsb.edu/primer.html.
[17] Peter Ward, "What May Become of Homo Sapiens," *Scientific American*, November 1, 2012,
https://www.scientificamerican.com/article/what-may-become-of-homo-sapiens/#.

transforming our environment so that it adapts to us rather than the other way around. We have essentially no more natural selection to shape human characteristics, because changes to our genes no longer have much effect on survival. In developed countries, virtually everyone survives, including individuals that might die if they had to fend for themselves out in the wild. In developing countries many babies die in childbirth or in early childhood, but mostly from disease or poor health care—nothing that is affected by their individual evolutionary fitness. Selection now is mostly driven by women choosing which men to mate with.

Eugenics

We could give evolution a boost, and improve human nature by changing the genotype. This could be done with traditional breeding practices, or with new genome-editing tools such as CRISPR.

Differentiation of dogs shows how effective traditional breeding can be. Look how different the dog breeds are from one another—if you didn't know, you'd think a Chihuahua and Saint Bernard are different species of animals. But it takes just a few hundred years to create a new breed by controlling which dogs mate with which others. It's artificial selection, akin to natural selection. You keep picking and breeding the dogs who have the characteristics you want.

Dogs evolved from wolves between 20,000 and 40,000 years ago.[18] The two species differ greatly in psychology: wolves are wild and independent; dogs are tame and subservient. Dog breeds also have different psychologies: my dog, a Kuvasz, is very independent and bridles at too much control, but a border collie revels in following complex orders to a tee. We could fix our human psychology by controlled breeding, but we should not do this for a variety of reasons.

Breeding takes effect through genes, which are part of the dog's DNA. In each generation, the genes of the father and mother are mixed together, the

---

[18] "Domestication of the Dog," in *Wikipedia*,
https://en.wikipedia.org/wiki/Domestication_of_the_dog.

results checked, and then the dogs with the most desirable traits are bred for the next generation, to try to create a mixture of genes that results in a dog a bit closer to the ideal in each generation.

We can theoretically achieve the same result by editing the dog genes with genetic engineering technology such as CRISPR. We could change dog characteristics by directly editing the genes, if we knew exactly what changes to make in the genes. But we are a long way from knowing which genetic changes to make to cause the results we want. We might need to make a large number of coordinated changes to achieve a relatively simple result such as changing hair color.

And we could do the same thing with humans. We seem to have started down that path—in 2018 a scientist in China modified genes in two babies born in China, in an effort to reduce the risk of HIV infection.[19] Changing human psychology would be much more difficult—we're a million miles away from knowing which genes to change. If we want to make people less tribal and nationalistic now, we'd have to do it by breeding, not by direct genetic manipulation.

Eugenics was all the rage about a hundred years ago.[20] By encouraging higher rates of reproduction among people with desirable traits, and sterilizing or discouraging the reproduction of those with less desirable traits, proponents thought the human species could be improved. In the early part of the 20th century, many countries enacted eugenics policies, including genetics screenings, birth control, promoting differential birth rates, marriage restrictions, segregation, compulsory sterilization, forced abortions, and forced pregnancies. These practices were "widespread and thoroughly respectable."[21] The belief that the Nordic or Aryan white race was

---

[19] David Cyranoski, "The CRISPR-Baby Scandal: What's next for Human Gene-Editing," *Nature* 556 (2019): 440–42, https://doi.org/10.1038/d41586-019-00673-1.

[20] Teryn Bouche and Laura Rivard, "America's Hidden History: The Eugenics Movement," *Scitable*, September 18, 2014, https://www.nature.com/scitable/forums/genetics-generation/america-s-hidden-history-the-eugenics-movement-123919444/.

[21] Tony Judt, *Postwar: A History of Europe since 1945* (Penguin Books, 2005), 368.

superior to the other races was purportedly supported by intelligence tests given to immigrants on Ellis Island starting around 1912. It was eventually realized that those tests were invalid because they did not adequately control for cultural, educational, or linguistic differences in the test subjects.

The Nazis embraced "scientific" racism. The Holocaust was an exercise in eugenics. In reaction, most western democracies repudiated eugenics, though in Scandinavia, a eugenics-based forced-sterilization program continued until 1976.

Reproductive rights, which preclude the practice of eugenics, have been recognized world-wide, starting in the late 1960s. The UN Declaration on Social Progress and Development declares that "parents have the exclusive right to determine freely and responsibly the number and spacing of their children."[22] Any eugenics program undertaken now would therefore need to be voluntary.

It is unlikely that the human race will, without our intervention in this process, evolve its psychology to better suit us to our role as stewards of Life on Earth. It hasn't happened during several thousand years of civilization.

Other ways to improve human nature

The moral code by which we live in society is an attempt to corral human nature, restraining its antisocial aspects and encouraging traits that help us work together. The code is embodied in our formal laws and regulations, but the informal code of mores is just as important. Laws and mores vary from place to place.

One of the functions of religion is to guide human nature so that we can live together well. Religions all include codes of morality, and laws and mores are usually based on the religious codes of the local inhabitants.

---

[22] "U.N. General Assembly Resolution 2542, Declaration on Social Progress and Development" (United Nations, December 11, 1969), https://www.ohchr.org/sites/default/files/Documents/ProfessionalInterest/progres s.pdf.

Some behaviors that are toxic to proper functioning of society are universally proscribed: murder and stealing, for example. Religious-based morality is failing us now because it is too human-centered, and too parochial, promoting us-vs-them thinking. It doesn't provide a basis for dealing with our environmental crisis. But, in the public sphere, we can evolve our laws and mores without support from religion.

And in the private sphere, individuals can work to change their own human nature. This chapter is an invitation to do that. It's hard work, and requires a thoughtful bent. It's much easier to accept one of the pre-packaged set of values and mores offered by organized religions than it is to develop one's own and then to live by them.

Humanism is speciesist

I used to think of myself as a secular humanist. "Secular" comes from the Latin "saeculum," which means a generation or age of humans. This is the sense in "novus ordo saeclorum," a "new order of the ages," the strange, mystical motto on the Great Seal of the United States, reproduced on the back side of the US one-dollar bill. The new order referred to is the new American era, which commenced in 1776, the date shown in roman numerals at the bottom of the great seals' pyramid. During the Middle Ages, the term "secular" took on the meaning "belonging to the world and its affairs, as opposed to the church and religion," which is its primary meaning today.

Humanism has a long history. According to an ancient Greek sophist named Protagoras, who lived in the fifth century BCE, "man is the measure of all things." (This statement is not sexist—the Greek word used for man, "anthropos," is gender-neutral.) During the Renaissance, when classical learning and philosophy were being re-discovered following the Church's domination of spiritual matters during the Middle Ages, scholars were impressed with the rationalism and rejection of the supernatural in ancient Greek and Roman philosophy. These ideas were taken up during the Enlightenment, which provided a lot of the intellectual material for the American revolution. Religious freedom, including the freedom to not believe in any religion, was therefore guaranteed in the US and France following their respective revolutions.

We tend not to notice that the term "humanism," like many of the other catch phrases that include the word "human," is speciesist since it is chauvinistically centered on human beings, implicitly omitting other forms of life from the moral universe. "Human rights are rights inherent to all human beings, regardless of race, sex, nationality, ethnicity, language, religion, or any other status."[23] This concept needs to be expanded beyond humans if we are to give moral status to other animals. That is why I am no longer a secular humanist. I consider myself a secular zoist, "secular" because I reject supernatural explanations like God, a "zoist" because I believe in life as the source of values.

Human divisions

We divide ourselves in so many ways. First, there is the human/other-animal divide just discussed. Second, we identify very strongly with the nation in which we were born and live. There are myriad further divisions and partisan associations—sports teams, religions, race, culture, gender, political party, sexual orientation, etc.

We're psychologically adapted to focus on these divisions because of our hunter-gatherer background. The most dangerous thing a person could encounter walking through the forest a hundred thousand years ago was not a wild animal, but another human from a different band or tribe. Groups of our ancestors fought one another, competing for territory and resources. This bonding with one's own group and fear and hostility toward other groups was built into our human brains during the eons when we lived as hunter-gatherers.

That psychology serves us badly now. It's a part of our human nature that we must fight as we redefine our ethics to deal with our new power over Life on Earth.

---

[23] "Human Rights" (United Nations), https://www.un.org/en/global-issues/human-rights.

Near and far

Jesus directed us to "Love thy neighbor as thyself."[24] When I was younger, I assumed this meant that we should all treat others the way we would like to be treated. I assumed that it applied to everyone. But the word "neighbor," ("πλησίον" in the original ancient Greek text), is based on the adjective for "near." Interpreted literally, it would mean that you should love the people nearest to you, but not necessarily those far away. The point here is not about what Jesus actually meant when he said this. I'm pointing out that even one of the most widely accepted statements that we should love other people contains, when read literally, a distinction between nearby and far-away people, a distinction we carry into our actual ethical practice.

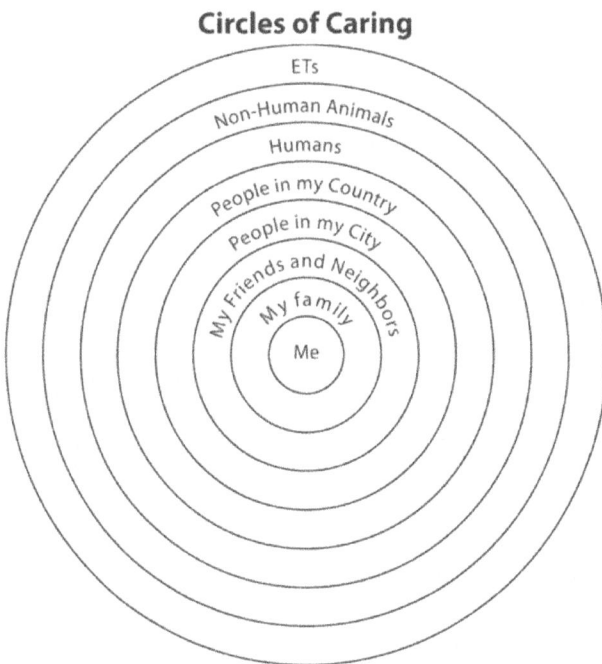

Circles of Caring

ETs
Non-Human Animals
Humans
People in my Country
People in my City
My Friends and Neighbors
My family
Me

Figure 1-1: *Circles of Caring*

The nearness need not be physical—family members and friends are "near" by blood, even if they live across the country. Each of us sits at the center

---

of a "circle of caring," where we come first—selfishness cannot be entirely banished—then our friends and family, the people who are most like us or near to us in some other way, then other people far away, then non-human animals. The Circles of Caring diagram highlights physical distances as the differentiator, but psychological and emotional differentiators provide other dimensions to the spheres: people we know have more moral weight for us than those we don't know; people like us get more weight, and humans get more weight than other animals.

There's a gradient: we care less about those in more distant circles, farther away from us, than we do about those near to us. The gradient is pretty strong; we care almost nothing for some poor soul living a miserable life on 50¢ per day in Sierra Leone. We insulate ourselves from knowing too much about such people because it makes us feel guilty. The five dollars we spend for a fancy coffee at Starbucks would significantly brighten the lives of many of the world's poor for a week or two; it only brightens our lives a little bit for an hour. If a poor Sierra Leonian appeared magically by my side, she would easily persuade me to give her the $5 rather than spending it at Starbucks, but she doesn't appear and I don't spend much time thinking about her.

Many of us care about non-human animals more than our espoused ethics would suggest we should, for example when the animals are our pets. My dog gets decent medical care, better than the bottom 50% of humans worldwide. Some of the best ads to raise money for environmental organizations depict "charismatic megafauna" such as wolves and polar bears. The Humane Society is very successful raising money with ads showing mistreated dogs and cats. A mistreated cat in our town gets more sympathy than a mistreated human far away.

Time is another dimension with its own gradient of caring. We care much less about a future person who will live in the same place we live in now than we do about our neighbor now. How much should we factor into our decisions their effect on someone a hundred years from now? A thousand years? This issue of how much we should care about and value future humans (and non-human animals) is very important to climate ethics, and will be discussed in much more detail in Chapter 3.

There's another gradient, kinship. I'm using this word in a very general way, to include not only family relationships, but racial and cultural kinship as well. We care less about people who are different from us, so we discount their value.

We have ethical gradients in time, space, kinship, and other dimensions by which we discount the effects of our decisions on humans and other animals. Even though it's not what he actually said, I've always interpreted Jesus' admonition to love my neighbor as myself to mean that we should love far-away people as much as our neighbor and ourselves. The zero gradient should apply to the future as well; we should care as much about a person who will live in a thousand years as we do about our friends now. For other species, the gradient should be determined by how sentient they are. A wolf deserves less moral consideration than a human, but more than a millipede.

Climate change brings out these ethical questions. Up until now, we could pretty much ignore them, but they have become important. Most of our moral decisions deal with people near us in space and time. We don't worry about the impacts of our actions on folks who will live in Kenya in five hundred years because, for most of our actions, there are no such impacts. But the climate change we are causing now will impact everyone on Earth for hundreds of years to come. Only in the recent past has the human race had the technological power to change our planet to this degree.

Only now do we have to face these new and difficult ethical questions. Being altruistic enough to properly deal with climate change requires reducing the time, space, kinship, and species gradients, so that we factor in impacts on faraway people and other animals in our decision-making. But such altruism is not in our human nature. Our nature prompts us to focus on the here and now, and to focus on ourselves, family, friends, and tribe—the central parts of the circle of caring.

Competition versus cooperation

Sports and business are related in our minds because they are both competitive. In the New York Times, the sports section is the last half of the business section. What's so great about competition? The theory is that it is the basis for our free-market society, which gives us all freedom and the chance to prosper economically. There is some truth to this, as will be discussed in Chapter 4. Another concept beloved by market economists is that, when you do something that produces value, like building a house, the value you created should belong to you and you alone; you shouldn't be forced to share that value with others. Forcing you to do so is "socialism."

But cooperation and competition are a yin-yang pair, inextricably bound. An entrepreneur who profits from developing a new technology is working within a support matrix, which nourishes her and provides essential services, like credit, a court system to enforce contracts, roads, a patent system allowing her exclusive rights to exploit her invention for a time, office buildings, treaties to facilitate exports, employees, stock exchanges to allow her to sell shares in her enterprise, etc. These services are provided by cooperation, often through government, not by competition.

Libertarians tend to ignore or discount the value of this matrix, and attribute all of the value created in building a business to the entrepreneur's efforts. This is one of the many points at which the far right touches the far left. German Socialist Karl Marx's labor theory of value attributes all economic value to the labor that created the value. According to his theory, business owners use capital to unfairly appropriate much of that value. Libertarians' analysis amounts to a labor theory of value for capitalists: they attribute the value of a business primarily to the efforts of the owner, the capitalist. These competing theories suggest different allocations of the profits from business: Marx thinks all such profits should belong to the worker, and the libertarians think they should all belong to the owner. There has of course been a major shift over the last hundred years: a much larger share of the profits is going to owners, greatly increasing income inequality.

The logical end of the libertarians' argument is that we could all be billionaires if we were smart enough to create wealth on a large scale. But the libertarians' model is wrong. Many folks with high incomes aren't rewarded for the value they produce; they are grabbing for themselves the value produced by others. Getting money—which, for most folks is getting resources to live a better life, but for the wealthy amounts to hoarding resources they will never be able to use—is, to a great degree, a zero-sum game. Our economic system generates a certain amount of wealth, and the competition is to get a larger share of it. What one person gets is taken away from others. This zero-sum aspect is most obvious with businessmen who make money by making deals, like real-estate magnates or corporate raiders. When they win, someone else loses.

How fair is it to reward so richly those who succeed by arrogating to themselves the economic value produced by others? Those who are good at winning the zero-sum game get a share of income wildly out of proportion to the economic value they produce. If everyone just made deals, there would no goods manufactured or services sold; there would be nothing to make deals about. Another reason we should distribute wealth more evenly is that the dealmakers and business owners benefit so much from the economic ecosystem we live in. They should contribute more to support that system, through taxes and by paying their workers better.

Individual vs. collective decisions

Which decisions should be made by individuals, and which should be made collectively? The assumption I'll make in discussing this question is that collective decisions will be made democratically. We can all agree to eschew totalitarian regimes and absolute monarchies.

Making all important economic decisions collectively is called communism. Under communism, the means of production, capital, factories, and companies, are collectively owned and managed by the government. Soviet attempts to build a strong, centrally planned economy have shown that it doesn't work well. Markets, a collective mechanism built on individual buy and sell decisions, are much better at efficiently allocating resources to produce the goods and services people want.

Markets provide a way to aggregate individual decisions into collective decisions, but there are a lot of problems with markets, some of which will be discussed in the chapter on economics. The biggest one is they consider everything from an economic point of view and don't adequately factor in non-economic values and concerns such as maintaining biodiversity or providing good health care for everyone. The second-biggest one is that it's not equitable because laissez-faire economics is largely a zero-sum game. Those who play the game well get a huge share of the wealth, and those who play it badly get nothing.

The political system provides another mechanism for making collective decisions, but it's badly broken in most countries. Does anyone think that our governments are implementing anything close to the will of the people? There are too many constraints and our politics have polarized into partisan zero-sum games.

Nonetheless, we agree that certain things must be decided collectively. For example, we consider actions such as murder to be criminal. We don't just prohibit these actions, but provide an elaborate criminal-justice system to administer penalties for those who commit them. Criminalization of certain behavior is based on a collective determination that certain actions are wrong because they are detrimental to the proper functioning of society.

Patriotism

Is our country really the best? Should we be loyal to our country and, if so, how far should that loyalty extend? And what is the philosophical basis for that loyalty? Is patriotism a good thing?

Most people didn't choose to live in the country where they reside; they were born there. There are about 281 million people living as immigrants in countries where they were not born.[25] This is about 3.5% of the world population of 8 billion. The US has a relatively high percentage of immigrants: 51 million,[26] just over 15 percent of the population. Immigrants might be more loyal to their new country because they made a choice and, in many cases, expended a great deal of effort, to be there. But why should the rest of us be loyal to a country, just because of the happenstance of having been born there?

---

[25] "World Migration Report 2022" (International Organization for Migration (IOM), 2021), 3, https://publications.iom.int/books/world-migration-report-2022.
[26] "World Migration Report 2022," 9.

Loyalty to the Dodgers if you live in Los Angeles, or to the Bruins if you're a UCLA alumnus, is expected, but fans are often attached to teams by default, not by explicit choice. Most fans don't choose teams by researching which team is best under the fans' criteria.

Loyalty to a country is like loyalty to a sports team – it's your team so you're loyal whether the team is good or not. Many Americans have never experienced life outside the US. Fifty percent of Americans have never owned a passport,[27] which means they have never been outside North America. Many Americans have no valid basis of comparison between the US and other countries. Their loyalty is based on belonging to the team, not on any determination that their team is the best.

The problem with patriotism is that it contributes to artificial factional barriers between people. Like all forms of team loyalty, it plays into the part of our human nature that wants to defend our band of hunter-gatherers from rival bands. This part of human nature is inappropriate for the modern world, where we don't live in bands of hunter-gatherers. The most important pressing problems of our time—climate change, especially—will be better solved by global cooperation than by competition among nation states.

I like to consider myself more a citizen of the Earth than a citizen of the United States. I choose to live in a particular part of the City of Los Angeles, in the State of California, United States of America, on Earth. But in a lot of ways I feel more European than American, strongly connected to France and Germany, even though I don't live there.

My wife, Benita, threatens to move to another country when she disagrees with our country's leadership, as she did during the Trump era. Such a move would make a political statement rejecting that leadership. But the statement would not be worth the financial cost of the move. I sometimes

---

[27] Lea Lane, "Percentage Of Americans Who Never Traveled Beyond The State Where They Were Born? A Surprise," *Forbes*, May 2, 2019, https://www.forbes.com/sites/lealane/2019/05/02/percentage-of-americans-who-never-traveled-beyond-the-state-where-they-were-born-a-surprise/?sh=203e48792898.

have the opposite impulse. Maybe I could do more good by living in a state with a much lower environmental consciousness. I could persuade them to do better, and my vote would count more in national elections. In the end, it seems to me that living in California, or in the US, is not a vote of approval for those political entities or their current political leadership. We live here because we have roots and jobs here, and because Los Angeles provides a lot of cultural resources that are valuable for us.

Patriotism is linked to nationalism, the notional opposite of globalism. There has been a recent dialectical swing from globalism to nationalism in many developed countries. Since World War II we've been increasing globalism by negotiating multi-party international treaties among many — and, in some cases, all — countries. These treaties become law in various ways and constrain the sovereignty of the State parties. In the last few years the pendulum has swung back. Nationalist parties in many developed countries want to reverse this process, putting national interests and culture ahead of international cooperation. This is like libertarianism for nations. The argument against it is similar to the argument against libertarianism: there are very important problems such as climate change, that require cooperation among nations to solve.

Money, money, money

Chapter 4 deals with the economics of climate change, but there is a moral and ethical underpinning which I will discuss here. The relationship between economics and ethics is fraught, because the two disciplines are fundamentally incompatible.

When we spend money it affects other people. We allocate physical resources that may be scarce, and we control, to a degree, how other people spend their time. Two million dollars can buy a person's working life in a relatively rich country: $25/hour × 2000 hours/year × 40 years = $2 million. A working life can be bought for much less than this in a developing country and, of course, some skilled working lives cost much more. If I buy an object that required 80,000 hours of labor (like a fancy yacht), I have caused the laborers to spend, in total, one working lifetime producing my object. Perhaps the person who built my yacht didn't really want to spend their lifetime that way; maybe that was the only work they could get. And maybe building the yacht is a bad use of resources for society as a whole. This same thing happens in a smaller way when I buy anything at all: I indirectly cause others to spend time and resources in a particular way. My purchase has ethical implications.

Jesus said that "it is easier for a camel to go through the eye of a needle than for a rich person to enter the kingdom of God."[28] Like Jesus, some on the political left are highly suspicious of wealth, and agree with a quote attributed to French novelist Honoré de Balzac: "behind every great fortune is an equally great crime." They feel that most rich people did something immoral to get their wealth. Just look at the robber barons from a century ago like James Fisk Jr., Jay Gould, and John Rockefeller. Catholics embrace this view in their tendency to feel guilty about wealth, and sometimes try to hide it from view.

Protestants and conservatives think wealth is a badge of honor, outward evidence of virtue. God rewards virtuous, thrifty, hard-working folks with prosperity. Poor people are mostly lazy, shiftless, slackers who deserve their poverty.

Another important split in attitudes towards money concerns whether those who make money owe anything back to society. The right-wing libertarian view is a version of Marx's labor theory of value recast for entrepreneurs: they alone are responsible for creating their wealth out of nothing, by their creativity and hard work. The logical conclusion of this view is that we could all be as rich as the billionaires Bill Gates or Elon Musk if only we had their talent and worked as hard. This is obviously nonsense: the economy doesn't produce nearly enough income for everyone to be a billionaire. This suggests the contrary conclusion: that making money is largely a zero-sum game. Those who are good at the game get the money at the expense of those who are not good at it.

How fair is it to reward so richly those who succeed by arrogating to themselves the economic value produced by others? Those who are good

---

[28] *Bible*, Matthew 19:24.

at winning the zero-sum game get a share of income wildly out of proportion to the economic value they produce. If everyone just made deals, which is the best way to make money quickly, there would no goods manufactured or services sold; there would be nothing to make deals about. Another reason we should distribute wealth more evenly is that the dealmakers and business owners benefit so much from the economic ecosystem we live in. They should contribute more to support that system, through taxes and by paying their workers better.

Rich people should care less about money because their basic needs are satisfied. American psychologist Abraham Maslow came up with a hierarchy of needs in 1943.[29] Under this theory, humans must meet their basic needs for food, shelter, and safety, before they can focus on higher-level needs such as social belonging, self-esteem, self-actualization, and transcendence.

In 2019, about 9.3% of the world population lived in extreme poverty, on less than USD $2.15 per day.[30] These folks need to focus their physical needs, the things they require just to stay alive. They barter for, or earn money to pay for, the resources they need. Those of us who live in developed countries tend to lose sight of how rich we are, compared to most people in developing countries. As of 2012, to have an income in the top one percent worldwide, you needed to earn USD $34,000 per year.[31] About half of US residents have sufficient income to be include in the top one percent worldwide.

But studies have shown that, once a person has enough money for the necessities, increased wealth adds only marginally to happiness and well-being. Why do so many of the very rich fight so hard to grab yet more

---

[29] "Maslow's Hierarchy of Needs," in *Wikipedia*, https://en.wikipedia.org/wiki/Maslow%27s_hierarchy_of_needs.

[30] "Poverty" (World Bank, November 30, 2022), https://www.worldbank.org/en/topic/poverty/overview.

[31] Hugo Gye, "America IS the 1%: You Need Just $34,000 Annual Income to Be in the Global Elite... and HALF the World's Richest People Live in the U.S." (Daily Mail, January 5, 2012), https://www.dailymail.co.uk/news/article-2082385/We-1--You-need-34k-income-global-elite--half-worlds-richest-live-U-S.html.

wealth for themselves? The difference in life satisfaction between billionaires and those with net worth of only a hundred million dollars must be very small, yet the top 1% worldwide have managed, between their economic activities and their political connections, to grab more and more of the world's resources for themselves. This extreme inequality warps our societies in many ways. The 1% have taken resources, in the part of the economy that's a zero-sum game, from others who need those resources much more than they do. Not only that, they use their wealth and political influence to continue to pass laws that increase their share of the wealth. This is very wrong, but they lean on the writings of Ayn Rand and their libertarian beliefs when they rationalize that, if everyone pursues his or her selfish interests, it will maximize happiness and well-being in the world at large.

The value of human life

Economists seek to maximize a quantity called "utility," which, in economics-speak, is a measure of satisfaction an individual gets from the consumption of goods or services. Utility correlates strongly with the price an individual is willing to pay for a good or service.

A more generalized, holistic, and humanized concept is "well-being," which includes having positive emotions and moods, the absence of negative emotions such as depression and anxiety, satisfaction with life, fulfillment and positive functioning.[32]

What economists would like us to do, though they don't often say so in so many words, is to use utility as a proxy for well-being. This requires the simplifying assumption that the economic side of life is of primary importance. Rich people tend to think of their wealth as their measure of success in life—the dollars are a way to keep score. This attitude trickles down to the rest of us, who rank ourselves by money and position. But if we ranked ourselves by well-being instead, the ranking would be quite different. Life satisfaction comes more from family and friends, self-esteem,

---

[32] "Well-Being Concepts" (Centers for Disease Control and Prevention, October 31, 2018), https://www.cdc.gov/hrqol/wellbeing.htm.

good mental health, and a chance to make a positive contribution to society than it does from money.

Another problem with using utility as a proxy for well-being is that a given amount of money matters a lot more to the poor than the rich. A gift of $1,000 would be a trifle for a billionaire, but could be life-changing for one of the billions of people living on $2/day or less. Adding another billion dollars to gross domestic product (GDP) increases utility, but our economic system is structured so that most of the benefits go the rich, and overall well-being increases very little as a result.

A friend, discussing global warming, suggested the way to find the best policy option would be to do a cost-benefit analysis. It's an attractive idea because the analysis can be done in a numerical way, using accepted economic principles. But the implicit assumption is that maximizing economic utility will provide the best solution to the problem, that utility is a proxy for well-being.

The legal system tends to use earning potential to value human life. Civil damages for killing a young surgeon who will probably make a million dollars a year for thirty years are likely to be much larger than damages for killing a person the same age who works taking orders at McDonald's. The US Environmental Protection Agency uses a figure of $7.4 million, regardless of the age, income, or other population characteristics of the affected population, to value a "statistical life."[33] They use this figure in cost-benefit analyses to value increased risk of death. A 1% risk increase for one person would be a cost of 1% of $7.4 million, or $74,000. There have been suggestions that it would be better to base the calculation of losses on human life years lost rather than lives lost, and this makes sense to me.

Ethics and economics are different spheres with incompatible rules. For this reason, we should be wary of deciding ethical questions with economic analysis. Deciding the proper response to the threats posed by climate change—the subject of this book—requires an overall ethical framework,

---

[33] "Mortality Risk Valuation" (US Environmental Protection Agency, March 30, 2022), https://www.epa.gov/environmental-economics/mortality-risk-valuation.

not an overall economic framework. Economics fit into the overall ethical analysis at certain points, but are not primary.

Are all lives worth living?

If I make a new mouse by breeding two mice, and then horribly torture the new mouse for a month before killing him, is there some value in his having been alive? He had a horrible life and never enjoyed it a bit, and his life wasn't useful to anyone else. He consumed resources that could have been used for something more beneficial. It seems to me that his life had a negative value—it would have been better if he had never come into existence.

But if my ethics are based on life, there surely must be some value in the creation of the mouse life, something to put on the other side of the scale when deciding whether the mouse's life was worth living. A normal laboratory mouse would live for a few months in a cage, and might enjoy that life, as long as he wasn't harmed. Is there an intrinsic value in the mouse having lived, on top of this potential enjoyment? I would say no, that there is no value in being alive, apart from the joy of living, and the value of the life to others.

The tortured-mouse question is an extreme version of the ethical question we ask about animal testing and raising animals for food. Those situations are somewhat different, though, because, in those cases, the animals' lives provide utility to humans, and the animals may not always have terrible lives. Chickens raised for slaughter and mice used for drug testing may have good days when they're enjoying themselves and glad to be alive. But, depending on the conditions in which they are kept, they may, in fact, have no such good days and would prefer to die rather than continuing their life under those conditions.

The same question could be asked for humans: are there some human lives that are so miserable and devoid of value to others that the humans in question would have been better off not being born? There may be some humans whose lives, because of poverty and oppression, are just as bad as a lab rat's life. But, aside from lives lived in terrible conditions, is there an intrinsic value to (human) life, apart from the enjoyment of life, and the value of that life to others?

This question comes into play when we consider what would be the ideal human population for our planet. If I had a knob that adjusted the annual change in worldwide population, up to a few percent per year, I would be able control the long-term population of the planet. What would be the ideal population? If I set the growth rate to seven percent per year, the population would double over the next 10 years, from 7.5 billion to 15 billion. If I set the growth rate to zero, the population would stay the same. What is the (moral) value of the extra 7.5 billion people we'd get by allowing the population to grow? Some of them would have happy and productive lives, which is a positive value. Some would have terrible lives of poverty and misery. Do we count those lives as negative values in the equation? Conversely, if our Earthling values are based on valuing life, as this chapter proposes, isn't the creation of more life a positive moral good?

My answer to this question would be that good lives have a positive value and bad lives a negative value. "Good" and "bad" are measured by the subjective quality of the life in question. A tortured mouse or person is a moral negative, and a happy mouse or person is positive. This answer implies that the optimal human population level would be the level that maximizes the worldwide total human (and other animal) happiness minus the worldwide total unhappiness. That formula suggest we should ramp up the population as high as it can go, if everyone has good living conditions.

As we will discuss in the chapter on sustainability, the big problem with a big population growth is that it's not sustainable—it uses Earth resources we can't replenish and will lead to degradation of everyone's lives. And it harms non-human life. We're in the middle of a mass extinction caused by our overuse of Earth's resources.[34]

---

[34] Elizabeth Kolbert, *The Sixth Extinction* (Henry Holt, 2014).

What should we do?

What I've tried to do, on the work side of my life, is to find something that needs doing and do it. Since I'm not independently wealthy, an important part of the equation is finding a way to get paid to do it, usually by starting a small business. Right now, the "thing that needs doing" is to fight climate change. I do this as an environmental public-interest lawyer, and make a living litigating against sprawl housing and warehouse projects in Southern California; our clients are environmental non-profits and citizen's groups.

When I was in my twenties, I designed one of the earliest commercial digital synthesizers and started a company to manufacture and sell them. In my forties I founded a company to sell educational CD-ROMs on the newly invented Internet. I'm proud of the contributions I made in both these projects. In between these efforts I worked as a software engineer, focusing on audio and financial applications.

I am lucky to have two sets of skills—software engineering and litigation— which can earn me a middle-class living. I could have made a lot more money than I did, using either of these skill sets, if that had been my main goal. The highest-flying big-law-firm partners I litigate against earn ten times as much as I do every year. But my consolation is that, by working for an environmental non-profit, I'm doing some good for the world rather than just helping big companies make more money. In my final years in software development, I was doing software-architecture consulting on payments systems for big New York banks. One of the drivers in my switch to environmental law was the thought that the main, big-picture, result of my work life was to help big banks make a little more money. That's not necessarily a bad thing, but it's not what I want my life to count for.

I am lucky to have been born into an affluent country, into a middle-class family that valued education. My education has given me the freedom to choose the work I do. Most people in this world have to devote all their work energies to surviving; the worldwide median per-capita income is just a few thousand dollars. But those of us who have been lucky enough to have a decent living should spend a significant portion of their life working for the common good.

Conclusion

There have been a number of proposals dealing with climate change to widen the scope of the measures to be taken beyond those specific to climate issues. For example, the Green New Deal, proposed by several members of the US Congress, calls for a more socialistic economy. They make a good case for the proposal, based on human rights and social justice.

But a more fundamental, ethical, shift is required. We have wiped out 69% of mammals, birds, fish, and reptiles since 1970.[35] If animals have moral standing, this is a holocaust. We should be taking their interests to heart, and we're clearly not.

In the Anthropocene, we humans manage the Earth, and we must manage it responsibly and sustainably, in a way that sustains all life, not just human life. There is an assumption among conservative politicians that the Earth will take care of itself, as it has for billions of years. Or perhaps that God manages the Earth as whole, so we don't need to get involved.

But, as we'll see in the next chapter, the Earth is not taking care of itself, and we're doing a miserable job managing it. We need to start taking effective action to stop climate change. If we don't, future humans will point back to this moment in time when humans irreversibly messed up the Earth. What we're currently doing to the Earth, if we don't stop it, will be the most important event in human history, since it will strongly harm the lives of future generations for at least hundreds of years, and will discard the majority of the rich biodiversity our planet has developed over millions of years.

---

[35] "Living Planet Report 2022" (World Wildlife Fund, October 13, 2022), https://livingplanet.panda.org/en-US/.

Chapter 2
Climate Change

The purpose of this chapter is to give the reader a broad general knowledge of climate change science, and to provide sources where the reader can find more detailed, in-depth information about climate change science.

Climate Change is the most significant ethical, political, and economic issue of our time. In my view it is more important than, say, World War II, because it will profoundly affect life on Earth for hundreds or thousands of years. When people a thousand years from now look back, they will say that this point in history—the early part of the 21st century—was when humans messed up the planet. As a result of our actions, the planet's climate became much less hospitable; we lost a great deal of coastal land to sea-level rise, and half Earth's species were driven to extinction. We have already done a lot of this damage, and continue to do more. What we do at this point in history will affect life on earth for at least a thousand years.

Global warming was identified as a significant environmental issue by 1965. A report entitled Restoring the Quality of Our Environment was submitted to Congress by President Johnson's Environmental Pollution Panel of the President's Science Advisory Committee in November of that year.[1] The report stated that the evidence clearly and conclusively established that the CO_2 content of the atmosphere had increased, and that the increase could be attributed to the burning of fossil fuels. If the increase continued, by the year 2000 there could be an increase in mean global temperature between 0.6 to 4.0°C.[2]

This forecast has turned out to be accurate—the increase in global temperature by the year 2000 was close to 0.5°C. It is currently over

---

[1] Environmental Pollution Panel, President's Science Advisory Committee, "Restoring the Quality of Our Environment. Report," November 1965, https://legacy-assets.eenews.net/open_files/assets/2019/01/11/document_cw_01.pdf.

[2] Environmental Pollution Panel, President's Science Advisory Committee, 121.

1.0°C.[3] Scientists have understood the basics of climate-change science for over fifty years. We've known for a long time what we should do to remedy the problem: stop burning fossil fuels. But we have not developed the political will to do this.

How science works

I sometimes see criticisms of climate science saying that the science is just a theory, that it hasn't been proven, and we should therefore discount it. This criticism is based on a misunderstanding of the way science works.

Using the scientific method, we devise hypotheses to explain the world, and then construct experiments to try to disprove them. Experiments should be reproduceable by others. Those hypotheses that survive attempts to falsify them become scientific truths, but they are always subject to disproof and refinement. For example, Newton's "law of gravity," states that the gravitational force of attraction between two objects is GMm/r^2, where G is the gravitational constant, M and m are the masses of the two objects, and r is the distance between them. It has worked well in practice since Newton proposed it in 1686. But Einstein's theory of general relativity, published early in the 20th century, showed that the law of gravity is really just an approximation. It's a very good approximation at the scale of normal objects and times we encounter in our lives, but it breaks down in extreme circumstances. Both Newton's law of gravity and Einstein's theory of general relativity are scientific theories.

Such theories are the end result of science. They provide the best explanations we have for physical reality, but they are always provisional. There's no such thing as a scientific law that is beyond disproof. Even the law of gravity is just a theory, though it's a theory that has shown itself to be accurate and useful for hundreds of years.

Science is an open process, to which anyone may contribute. Results of experiments are usually publicly reported in scientific journals, and the

---

[3] NASA Global Climate Change, "Global Surface Temperature | NASA Global Climate Change," Climate Change: Vital Signs of the Planet, https://climate.nasa.gov/vital-signs/global-temperature.

reports should contain enough information to allow an interested party to repeat the experiments to verify the results. Likewise, the development of theories from experimental results is a process that is documented in scientific journals.

Climate science is interdisciplinary and complicated. Some of its conclusions are well established among the scientific community, such as the "fact" that global mean surface temperature on Earth has increased by more than 1°C since 1900. Other conclusions are more tenuous, such as the theory that climate change has strongly contributed to recent droughts in California. The IPCC reports (discussed below) deal with this by annotating each conclusion with its degree of certainty, ranging from "extremely unlikely" (between 0 and 5% probability) through "likely" (between 66% and 100%) to "virtually certain" (99%-100%).[4]

The degree of certainty for scientific conclusions is similar in concept to the standard of proof in the legal realm. In both cases, we want "proof," which, colloquially, means that the conclusion is certain to be true. But there is no absolute certainty in either realm. For most criminal convictions in the US, the jury has to find the defendant guilty "beyond a reasonable doubt." But the standard of proof in most civil cases is "preponderance of the evidence," which just means "more likely than not." So when we say in a legal context that a particular fact has been proven, we might just mean that someone has made a showing that it has a 51% chance of being true. When we say that a scientific theory has been proven, we usually mean that it hasn't yet been disproven by experiment. Newton's theory of gravity was not disproven during the first three hundred years after it was proposed. It exactly fit all the experimental evidence during that period. It wasn't until the early 20th century that experiments showed that it was an approximation, and inaccurate in some cases at very large and very small scales.

How do we deal with uncertainty about climate change? Climate obstructionists argue that we shouldn't harm our businesses by taking drastic steps to mitigate climate impacts when the science isn't certain. This

---

[4] IPCC WGI, "Climate Change 2021: The Physical Science Basis" (IPCC WGI, 2021), 4, fn. 4 https://www.ipcc.ch/report/ar6/wg1/.

is a reverse version of the precautionary principle, which says "when an activity raises threats of harm to the environment or human health, precautionary measures should be taken even if some cause and effect relationships are not fully established scientifically."[5] The precautionary principle, applied to climate change, says, in effect, "if we're uncertain how much climate change will harm Life on Earth, we should take precautionary steps now to limit its impact rather than waiting for a full understanding of those harms."

But the climate-change big picture is not uncertain. Human activities are increasing the concentration of greenhouse gases in Earth's atmosphere, resulting in a gradual warming of the planet. The primary cause is the burning of coal, natural gas, and oil. These facts are established with a very high degree of confidence, even if, as we shall see below, many of the details are known with less certainty.

There is a 97% consensus among climate scientists concerning the big-picture conclusions in the previous paragraph.[6] Nonetheless, there are a few credentialled scientists who dispute these conclusions. Their arguments sometimes seem to make sense, and are difficult for laypeople to counter. Many of these arguments are listed and countered on the Skeptical Science Web site.[7]

How climate change works

Sources of information

The basics of climate change are simple; the details are very complex. The science is interdisciplinary, so it takes a wide range of knowledge to

---

[5] Mary Stevens, "The Precautionary Principle in the International Arena," *Sustainable Development Law & Policy* 2, no. 2 (2002): 6, https://digitalcommons.wcl.american.edu/cgi/viewcontent.cgi?referer=&httpsredir=1&article=1278&context=sdlp.

[6] John Cook et al., "Consensus on Consensus: A Synthesis of Consensus Estimates on Human-Caused Global Warming," *Environmental Research Letters* 11, no. 4 (April 2016): 048002, https://doi.org/10.1088/1748-9326/11/4/048002.

[7] "Arguments from Global Warming Skeptics and What the Science Really Says," Skeptical Science, https://skepticalscience.com/argument.php.

understand it well. I'll cover the basics here, but refer the reader to textbooks and other definitive sources for more depth.

Some of the best textbooks on climate-change science are:

- *Climate Change*, by Edmond A. Mathez and Jason E. Smerdon:[8] An in-depth textbook with a lot of scientific detail;
- *Global Warming, the Complete Briefing*, by John Houghton:[9] A fairly technical summary; and
- *Introduction to Modern Climate Change*, by Andrew Dessler:[10] A more readable, basic introduction for the layperson.

The most definitive source is the Climate Change 2014 Synthesis Report (AR5 Synthesis Report),[11] which was issued in 2015 by the United Nation Environmental Program's Intergovernmental Panel on Climate Change (IPCC). The report was developed by thousands of top climate scientists and approved by the governments of every country on earth. As of this writing, the three important Working-Group reports have been completed in the sixth cycle, but the overall synthesis report, summarizing the current situation has been delayed until mid-2023.

The Working Group I (WGI) report[12] covers the physical-science basis for climate change. It's a very technical report, and is currently the most definitive source of expertise on the physical-science aspects of climate change. Even though it's almost 4,000 pages long, it's a summary report, and cites thousands of scientific sources, which provide more detail in support of the report's conclusions. Similarly, the Working Group II (WGII)

---

[8] Edmon A. Mathez and Jason E. Smerdon, *Climate Change: The Science of Global Warming* (New York: Columbia University Press, 2018).

[9] John Houghton, *Global Warming - The Complete Briefing*, 5th ed. (Cambridge, UK: Cambridge University Press, 2015).

[10] Andres Dessler, *Introduction to Modern Climate Change*, 2nd Ed. (Cambridge, UK: Cambridge University Press, 2016).

[11] IPCC, "AR5 Synthesis Report: Climate Change 2014," 2015, https://www.ipcc.ch/report/ar5/syr/.

[12] IPCC WGI, "Climate Change 2021: The Physical Science Basis." https://www.ipcc.ch/report/ar6/wg1/.

report[13] covers the Impacts, Adaptation, and Vulnerability to climate change, and the Working Group III report analyzes mitigation of climate change.[14]

The greenhouse effect

Hot objects emit light; this is called black-body radiation. The color of the emitted radiation changes with temperature. A piece of metal slowly heated with a blowtorch will get red-hot, then glow orange, then yellow, then blue, as the temperature increases. The same piece of metal at room temperature emits infrared light, invisible to the human eye, but measurable with instruments. No-touch thermometers, like those used to take temperatures for COVID-19 screening, work by measuring the frequency (color) of the invisible infrared light emitted by the human body.

A major factor in determining the temperature of Earth's atmosphere is the radiation balance, the balance between incoming and outgoing radiation. The vast majority of incoming radiation is from the sun. The outgoing radiation is mostly infrared black-body radiation from the surface of the earth. If the incoming radiation decreased, the atmosphere would cool down. If the outgoing radiation decreases, the temperature goes up.

That's what's causing global warming: a decrease in the infrared light radiated back into space. Greenhouse gases like carbon dioxide and methane absorb specific frequencies of infrared radiation, preventing its energy from being transmitted back out into space. Retaining the energy on Earth causes the temperature to go up. This greenhouse effect becomes stronger as concentrations of greenhouse gases in the atmosphere increase, pushing up Earth's global average temperature. This effect was laid out in an 1896 paper by a Nobel prize-winning Swedish chemist, Svante August Arrhenius.[15]

---

[13] IPCC WGII, "Climate Change 2022: Impacts, Adaptation and Vulnerability" (IPCC, 2022), https://www.ipcc.ch/report/ar6/wg2/.

[14] IPCC WGIII, "Climate Change 2022: Mitigation of Climate Change" (IPCC, 2022), https://www.ipcc.ch/report/ar6/wg3/.

[15] Svante Arrhenius, "On the Influence of Carbonic Acid in the Air upon the Temperature of the Ground," *Philosophical Magazine and Journal of Science*, 5, 41 (April 1896): 237–76, https://www.rsc.org/images/Arrhenius1896_tcm18-173546.pdf.

The greenhouse effect is not generally harmful; it is an important part of the way the atmosphere maintains temperatures in a range that allows for life on Earth. Without the greenhouse effect, the average temperature on Earth would be around −18°C, way below freezing.[16] What is harmful is humans' intensifying the greenhouse effect by decreasing the outgoing infrared radiation, which disturbs Earth's radiation balance, increasing the amount of heat retained within the atmosphere.

Greenhouse gases

Carbon Dioxide (CO_2) is the most important greenhouse gas. We add a huge amount—around 36 billion metric tons—of it to the atmosphere very year, mostly through combustion of fossil fuels. The main CO_2 absorption band is centered right where Earth's infrared emissions are most intense.[17] In addition to being the GHG whose atmospheric concentration has increased the most, CO_2 has the longest persistence in the atmosphere among major GHGs—hundreds of years. Indeed, 20 to 35% of CO_2 may remain in the atmosphere for 2,000 years and 7% may remain after 100,000 years.[18]

Methane, the second-most-important GHG, is the main ingredient in natural gas. It's a much more powerful GHG than CO_2, but it's emitted in much smaller quantities, mostly from agriculture and fossil-fuel energy production. There are also natural sources, such as wetlands. Methane's "radiative efficiency," i.e., the amount it reduces radiation of energy from the Earth into space, is 29 times that of CO_2[19], though methane's lifetime in the atmosphere is just 9.1 years.[20]

The Global Warming Potential (GWP) of each greenhouse gas indicates how much impact the gas has on global heating. By definition, CO_2's GWP is 1.0. Methane has a GWP of between 7.5 and 82.5—meaning that

---

[16] Mathez and Smerdon, *Climate Change: The Science of Global Warming*, 143.

[17] Mathez and Smerdon, 165.

[18] Mathez and Smerdon, 170–71.

[19] IPCC WGI, "Climate Change 2021: The Physical Science Basis," 1012.

[20] IPCC WGI, 701.

methane is between 7.5 and 82.5 times more powerful than CO_2 in its retention of heat within the atmosphere, depending on the timeframe considered.[21] The figure of 82.5 is for a 20-year timeframe; 7.5 is for a 100-year timeframe. The difference results from methane's short life in the atmosphere. A ton of methane emitted today will have much more impact right now than a ton of CO_2 but most of it will be removed from the atmosphere within a few decades, so its long-term GWP is much less than its short-term GWP.

There are several other gases that, together, block about as much outgoing infrared radiation as methane. They have high GWPs, but exist in the atmosphere in very small concentrations. Nitrous oxide (NO_2) contributes about 7% of global warming.[22] Its 20-year GWP is 273.[23] Some NO_2 emissions come from the microbial breakdown of soils, both naturally, and in agriculture. The anthropogenic emissions could be reduced by better farming practices. Note that GHG emissions are sometimes referred to as "carbon emissions," but NO_2 is a good example of a GHG that does not contain carbon.

Hydrofluorocarbons (HFCs), chlorofluorocarbons (CFCs), and perfluorochemicals (PFCs) are GHGs with very high GWPs—between 2693 and 5301 on a 20-year basis. Their impact now is relatively small because their concentrations in the atmosphere are low.

The quantities of mixed gases are often combined pro-rata based on their GWP into a single number, usually expressed in units of $MTCO_2e$—metric tonnes of CO_2 equivalent. A "tonne" is a metric ton, 1000 kg, about 10% larger than a US or British ton. For example, a new housing project that results in additional emissions of 75 tonnes of CO_2, plus 2 tonnes of methane may be characterized, based on 20-year GWPs, as emitting 75 + (2 × 82.5) = 240 $MTCO_2e$.

---

[21] IPCC WGI, 1017.

[22] Houghton, *Global Warming - The Complete Briefing*, 278.

[23] IPCC WGI, "Climate Change 2021: The Physical Science Basis," 1017.

Water vapor is also a very important GHG that absorbs and scatters infrared radiation. There is a great deal of it in the atmosphere, but it doesn't come mainly from emissions by humans, like the other GHGs. Oceans, lakes, and rivers evaporate in the upward part of the cycle, and rain removes the water vapor in the downward part. Water vapor is mostly important as a feedback, discussed below: increasing global temperatures result in more water vapor in the atmosphere, which increases the greenhouse effect, further warming the Earth.

Higher temperatures correlate with increased GHG concentrations

The following graph from the IPCC AR5 Synthesis Report shows that global average temperatures had increased almost 1°C between 1900 and 2021. The increase is currently about 1.3°C.

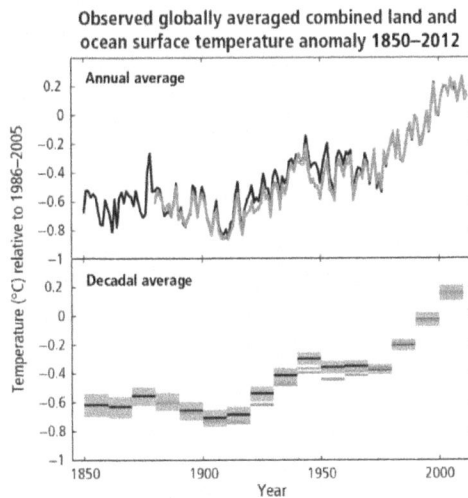

Figure 2-1: *Global temperature increases since 1850*[24]

---

[24] Figure 1.1 from IPCC, 2014: Topic 1 – Observed Changes and their Causes. In: *Climate Change 2014: Synthesis Report. Contribution of Working Groups I, II and III to the Fifth Assessment Report of the Intergovernmental Panel on Climate Change [Core Writing Team, R.K. Pachauri and L.A. Meyer (eds.)]. IPCC, Geneva, Switzerland, 151 pp.*

The following graph, based on data from the Carbon Dioxide Information Analysis Center, shows the history of GHG emissions in the industrial age.

The correlation between the emissions and the temperature increase is obvious. Correlation is not causation, but we know the causal mechanism in this case: the greenhouse effect.

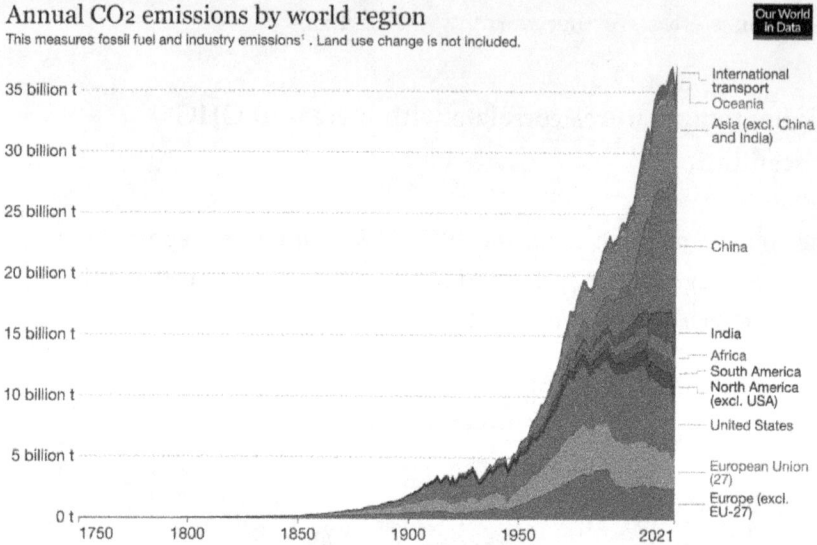

Annual CO₂ emissions by world region
This measures fossil fuel and industry emissions[1] . Land use change is not included.

Source: Our World in Data based on the Global Carbon Project (2022)   OurWorldInData.org/co2-and-other-greenhouse-gas-emissions · CC BY

1. **Fossil emissions**: Fossil emissions measure the quantity of carbon dioxide (CO_2) emitted from the burning of fossil fuels, and directly from industrial processes such as cement and steel production. Fossil CO_2 includes emissions from coal, oil, gas, flaring, cement, steel, and other industrial processes. Fossil emissions do not include land use change, deforestation, soils, or vegetation.

Figure 2-2: *History of GHG Emissions*

There are many other factors that affect the climate, of course, and climate-deniers try to attribute the rise in global average temperatures to them. These factors include changes in the Earth's orbit around the Sun, and the 11-year solar sunspot cycle. Climate scientists have considered these alternative potential causes and, almost universally, agree that the major cause of the warming of the planet since 1870 has been the accumulation of greenhouse gases in the atmosphere. There is no significant controversy on this point among climate scientists.

The ocean and global heating

The ocean plays a huge role in global heating: about 93% of the global heating since 1955 has been absorbed by the ocean.[25] The ocean interacts with the atmosphere via the water cycle; water evaporates from the ocean, becomes atmospheric water vapor, then returns to the ocean as precipitation. The ocean also exchanges energy with the atmosphere through conduction of heat between the two.

The ocean, like the atmosphere, is a large, complex, dynamic system. There are many currents and flows, like the Gulf Stream, some of which may be disrupted by global heating. There are several large-scale "oscillations," chief among which is the El Niño-Southern Oscillation (ENSO), which is on an approximately 4-year cycle between El Niño and La Niña conditions.[26] Those conditions affect droughts and precipitation in parts of the world south of the mid-USA. There are more oscillations: the Arctic Oscillation, the North Atlantic Oscillation, the Pacific Decadal Oscillation and the Atlantic Multidecadal Oscillation.[27] These oscillations do not affect the total amount of global heating, which is controlled by Earth's energy balance, but they do affect the distribution of the impacts of global heating.

The carbon cycle

Photosynthesis is a chemical process by which plants convert CO_2 and water into oxygen and glucose, a sugar that the plant uses for energy. When the plant consumes glucose for growth, respiration reverses the process, producing CO_2 and water. Typically, plants end up releasing about half of the CO_2 they absorb during photosynthesis back into the atmosphere during respiration. The other half remains as a component of the organic material within the plant.

Large plants, such as trees, can sequester a lot of carbon in their trunks, branches, and roots, and can also, as they decay, generate organic matter

---

[25] Mathez and Smerdon, *Climate Change: The Science of Global Warming*, 271–72.

[26] Mathez and Smerdon, 81.

[27] Mathez and Smerdon, 91–95.

that becomes part of the soil, sequestering still more carbon. But such decay also releases a substantial amount of CO_2.

Albedo

Albedo is the percentage of light that is reflected from a given surface. Dark surfaces like asphalt have low albedos, and light surfaces like fresh snow have high albedos. Changes in albedo affects global heating because low-albedo surfaces absorb more solar radiation and heat up more. For example, replacing high-albedo sea ice in the Arctic with low-albedo open ocean increases heating.

Land-use change

Changes in land use have important global-heating consequences. The biggest land-use contributor to global heating is deforestation. Cutting down forests eliminates them as carbon sinks. If the cut material is burned or decomposes, it releases its CO_2 back into the atmosphere.

Deforestation is an especially large problem in tropical forests, such as those in the Amazon River valley. Tropical forests store more carbon than do temperate or boreal forests. Deforestation is proceeding rapidly in the Brazilian Amazon in spite of Brazil's recent pledge to end the practice by 2030.[28]

Albedo often changes when land use changes. For example, fields and pastures tend to have higher albedos than the dark green forests they replace, reducing the heat that's absorbed. This effect is also pronounced in more extreme latitudes where snow has a higher albedo in cleared areas than did the trees that were cleared.

Aerosols

Aerosols are small particles, typically less than 2 microns in diameter, suspended in the air. Aerosols include black carbon (soot) from urban uses

---

[28] BBC, "Brazil: Amazon Sees Worst Deforestation Levels in 15 Years" (BBC, November 19, 2021), https://www.bbc.com/news/world-latin-america-59341770.

and forest fires, organic aerosols from fossil-fuel and biofuel burning, sulfate, a pollution from fossil-fuel power plants, sea salt from the ocean, biological particles such as pollen, algae, bacteria and viruses, and mineral dusts, mainly from deserts.[29]

Aerosols have complex and imperfectly understood interactions with clouds and the atmosphere, but aerosols, taken together, probably reduce the impacts of global heating.

Contributions to global heating

The following diagram from IPCC AR6[30] shows the contributions to global heating from human activities:

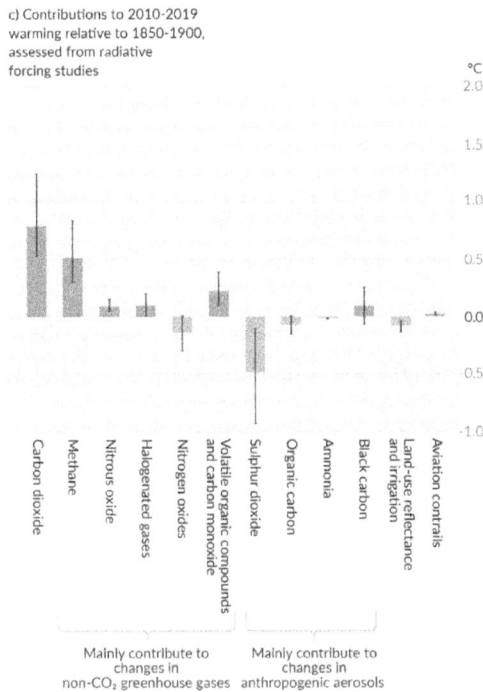

Figure 2-3: *Global heating contributions from various sources[31]*

---

[29] Mathez and Smerdon, *Climate Change: The Science of Global Warming,* 171.

[30] IPCC WGI, "Climate Change 2021: The Physical Science Basis," 7.

[31] Figures SPM2 (c) and SPM.8 from IPCC, 2021: Summary for Policymakers. In: *Climate Change 2021: The Physical Science Basis. Contribution of Working Group I to the*

Feedback loops

Feedbacks occur when the outputs of a system are routed back as inputs, affecting the system. A positive feedback increases the effect of the system, taking it away from equilibrium; a negative feedback decreases the system's effect, bringing it back toward a stable state.

An example of a negative feedback is the increase in photosynthesis that comes from increased atmospheric concentrations of CO_2. Plants absorb more CO_2 when the concentration is high; this absorption takes CO_2 out of the atmosphere, which feeds back to reduce their photosynthesis.

A climate-related example of a positive feedback is the melting of ice in the Arctic, replacing bright, high-albedo surfaces (ice) with dark, low-albedo surfaces (open ocean), which absorb more radiation. The replacement results in more heat energy being absorbed by the Earth, increasing global heating, which, in turn, leads to more ice melting.

Other positive feedback mechanisms are related to the release of methane as temperatures increase. Peat bogs on land, and methane clathrates, in which methane is trapped in crystalline ice structures in the ocean, could release large amounts of methane if they melt, leading to increased methane in the atmosphere, which will increase global heating, leading to more methane releases from peat bogs and methane clathrates.

Forest fires release a lot of CO_2 which, leads to more warming, and the increased temperature increases the number of forest fires. This is another positive feedback.

The interactions of clouds and water vapor with the atmosphere are complex, difficult to model, and difficult to understand. But the biggest water-vapor positive feedback is easy to understand: with increased

Sixth Assessment Report of the Intergovernmental Panel on Climate Change [Masson-Delmotte, V., P. Zhai, A. Pirani, S.L. Connors, C. Péan, S. Berger, N. Caud, Y. Chen, L. Goldfarb, M.I. Gomis, M. Huang, K. Leitzell, E. Lonnoy, J.B.R. Matthews, T.K. Maycock, T. Waterfield, O. Yelekçi, R. Yu, and B. Zhou (eds.)]. Cambridge University Press, Cambridge, United Kingdom and New York, NY, USA, pp. 3–32, doi:10.1017/9781009157896.001

temperature there is more evaporation and more water vapor in the atmosphere. Since water vapor is a potent GHG, this causes the atmosphere to trap more heat, increasing the temperature, and increasing evaporation.

Tipping Points

Most of the climate-related impacts we've discussed so far have been relatively linear, in the sense that small changes to the inputs of a system result in small changes to the system's outputs. Tipping points are thresholds in the climate system where small changes in the inputs could lead, immediately or in the longer term, to runaway feedbacks, causing large changes in the outputs.

Some potential tipping points are the disappearance of Arctic sea ice, the collapse of the Greenland Ice Sheet, the Collapse of the Western Antarctic Ice Sheet, the melting of permafrost in cold regions, the breakdown of ocean methane hydrates, and the disruption of ocean circulation patterns. We're currently thawing permafrost, and expect that, by the end of the century, permafrost area will decrease between 24% and 69%, resulting in the release of tens to hundreds of billions of tons of permafrost carbon as CO_2.[32]

A recent analysis suggests that the Amazon rainforest, the largest on earth, may be approaching a tipping point where it will all convert to grassland.[33] The loss of resilience in the Amazon is probably caused by a combination of deforestation—converting rainforest for human use for agriculture or pasturage—and climate change, another positive feedback. The Amazon is a huge carbon sink, so the loss of the rainforest would greatly reduce its removal of CO_2 from the atmosphere and cause more warming.

---

[32] "Special Report on the Ocean and Cryosphere in a Changing Climate" (IPCC, 2019), 18, https://www.ipcc.ch/srocc/.

[33] Chris A. Boulton, Timothy M Lenton, and Niklas Boers, "Pronounced Loss of Amazon Rainforest Resilience since the Early 2000s," *Nature Climate Change* 12, no. March 2022 (March 2022): 271–80, https://doi.org/10.1038/s41558-022-01287-8.

Environmental alarmists sometimes speak as though, if we don't manage to reduce our GHG emissions, climate change will result in the extinction of the human race. This is extremely unlikely, because humans are so adaptable. In the very unlikely event that climate change somehow killed 99% of humans, there would still be 70 million of us left, and we have enough technology, like air conditioning, to survive on a much hotter planet.

The one tipping point that could pose an existential threat to humans and most of life on Earth is called "Runaway Greenhouse," and is what the climate scientist James Hansen calls the "Venus syndrome."[34] The planet Venus probably had oceans at one time, but a runaway greenhouse effect evaporated all the water, resulting in a positive feedback where the greenhouse effect of more water vapor in the atmosphere increased global heating, eventually evaporating all the water from the oceans. The water vapor in the atmosphere dissociated into hydrogen and oxygen. The hydrogen boiled off into space and the oxygen combined with carbon, resulting in today's atmosphere on Venus, 97 percent carbon dioxide. There's another, more gradual, process called "Moist Greenhouse," by which the Earth could also lose its water to space relatively quickly.[35] It could be triggered by very high CO_2 concentrations, above 770 parts per million (p.p.m.). Another recent paper models stratocumulus clouds, concluding that at CO_2 concentrations above 1,200 p.p.m., those clouds will break up into cumulus clouds as a result of a local runaway greenhouse effect, eventually resulting in an additional 8-10°C of warming.[36]

Nobody suggests we're close to the tipping point that would trigger these runaway greenhouse scenarios, but, in truth, we don't know when that tipping point will be reached, or any other tipping point, for that matter. For example, the ice sheet covering Greenland contains enough ice that, if it were to melt, it would raise the level of the oceans, world-wide, by

---

[34] James Hansen, *Storms of My Grandchildren* (Bloomsbury USA, 2009), 223–36.

[35] Max Popp, Hauke Schmidt, and Joche Marotzke, "Transition to a Moist Greenhouse with CO2 and Solar Forcing," *Nature Communications* 7, no. 10627 (February 9, 2016), https://doi.org/10.1038/ncomms10627.

[36] Tapio Schneider, Colleen M. Kaul, and Kyle G. Pressel, "Possible Climate Transitions from Breakup of Stratocumulus Decks under Greenhouse Warming," *Nature Geoscience* 12 (March 2019): 163–67, https://doi.org/10.1038/s41561-019-0310-1.

about 7 meters. The mass of the Greenland ice sheet has dropped by more than 4,700 Gt since 2002.[37] This is enough to raise the global sea level by 1.2 cm. There is a feedback loop that may result in a tipping point leading to an irreversible loss: temperature is higher at lower altitudes. As ice is lost, the upper parts of the glacier lose altitude, and are subject to higher temperatures, causing them to melt even more quickly.[38] We don't know exactly where the tipping point is, but if we reach it we'll have committed the Earth to an unstoppable 7-meter sea level rise.

Earth's climate history

Even though we don't have direct measurements of temperature and other climate data going back more than a few centuries, we can use other scientific evidence to estimate climatic conditions millions of years in the past. We have a 70-million-year record of sediments accumulated on the bottom of the oceans, which allow us to estimate sea temperatures throughout that period.[39] And the ice sheets covering most of Greenland and Antarctica have been built up over hundreds of thousands of years as snow fell and compacted the layers of snow below. They contain not only ice but bubbles of air from the time of the snowfall, which may be measured to estimate GHG concentrations in the distant past.

In both Greenland and Antarctica scientists have drilled down deep and extracted ice cores containing a continuous record of climatic variables. Antarctic cores go back about 800,000 years; Greenland cores go back about 100,000 years. We can directly measure the CO_2 content in their bubbles, and the proportions of oxygen isotopes in the ice gives us a good proxy for temperature at the time the snow fell.[40] For the last 800,000

[37] Polar Portal, "Mass and Height Change, Greenland Ice Sheet," http://polarportal.dk/en/greenland/mass-and-height-change/.

[38] Mathez and Smerdon, *Climate Change: The Science of Global Warming*, 317.

[39] Dessler, *Introduction to Modern Climate Change*, 30–31.

[40] Matthez and Smerdon, *Climate Change: The Science of Global Warming*, 193–224 has a detailed account of Earth's past climate, from which the graph just below was taken.

years, the global temperature has closely tracked CO₂ concentration in the atmosphere:

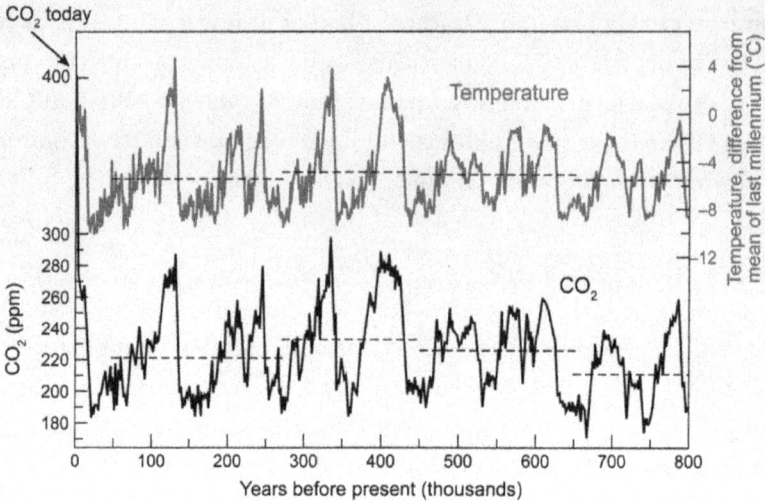

Figure 2-4: *CO₂ and temperature during the last 800,000 years*[41]

The above graph[42] shows ice ages alternating with warmer periods during that interval, with a period of between 41,000 and 100,000 years. There have been many such large changes in climate during the last 5 million years.

The current CO₂ level of 412 p.p.m. is higher than the maximum concentration in the last two million years.[43] Though there have been many large swings in the Earth's climate in the last 50 million years, the temperature has never increased as quickly as it's increasing today. Human interference with the climate is a much stronger driver of change than the Earth has ever seen.

Let's review the planet's climate history. Here's a list of major events in climate and human history:

---

[41] From Climate Change: *The Science of Global Warming* by Edmon A.Mathez and Jason E. Smerdon. Copyright © 2018 Edmon A. Mathez and Jason E. Smerdon. Reprinted with permission of Columbia University Press.

[42] Mathez and Smerdon, *Climate Change: The Science of Global Warming*, 211.

[43] IPCC WGI, "Climate Change 2021: The Physical Science Basis," 8.

Years Ago	Event
13,800,000,000	Big Bang - start of the universe
4,543,000,000	Creation of the Earth
3,500,000,000	First life on Earth
541,000,000	Start of Cambrian period
450,000,000	Ordovician-Silurian extinction events
251,900,000	Permian-Triassic extinction event
201,300,000	Triassic-Jurassic extinction event
66,000,000	Cretaceous-Paleogene extinction event
5,330,000	Start of Pliocene epoch
2,580,000	Start of Pleistocene Epoch - ice age
2,000,000	Genus homo evolved from apes
800,000	Start of 100K year glaciation cycles
200,000	Homo sapiens
125,000	Eemian interglacial
100,000	Humans develop language and complex reasoning
25,000	Last glacial maximum (LGM)
21,000	Peak of most recent glaciation
12,500	Humans invent agriculture
11,700	Start of Holocene Epoch
900	Little ice age

The first major event, which occurred about 14 billion years ago, is the "big bang," where the universe came into being, with all its matter and energy starting out compressed into a tiny "singularity," which expanded, quickly at first, into the universe we have today. The universe is still expanding.

Earth was formed about 4.5 billion years ago. The first, single-celled, life appeared about 3.5 billion years ago. Multicellular life had to wait for the Cambrian period, about 541 million years ago.

Between the Cambrian period and the start of the Pliocene epoch about 5 million years ago there were four major extinction events. The extinction at the end of the Ordovician killed trilobites and echinoderms and such, and was at least partly caused by a change in climate resulting in glaciation.[44] The End-Permian extinction event killed approximately 57% of biological families, 83% of genera, 81% of marine species, and 70% of terrestrial vertebrate species. It was caused by CO_2 emissions, temperature increases and ocean acidification. The extinction at the end of the Triassic, possibly caused by volcanic eruptions emitting CO_2, resulting in global heating, killed around 30% of marine species. The extinction we're most familiar with today is the one at the end of the Cretaceous, about 65 million years ago, when a large asteroid struck Yucatan and hurled enough dust into the atmosphere to diminish light levels planet-wide. About three-quarters of plant and animal species, including dinosaurs, perished in this extinction. All of these major extinction events were climate-related.

During most of the Pliocene epoch, which occurred between 5.3 and 2.6 million years ago, the global mean temperature was around 2.5°C higher than today, sea levels were about 20 meters higher than today, and CO_2 levels in the atmosphere were about the same as today.[45] We don't understand exactly why temperatures were so much higher than they are today, even though conditions were in many ways similar.

The Pleistocene began about 2.6 million years ago, and ushered in an epoch of ice ages. Until about 800,000 years ago they occurred every 41,000 years. After that, the periodicity changed to 100,000 years, each period containing a glacial phase lasting between 80,000 and 90,000 years, followed by an interglacial thawing lasting about 10,000 years.[46] These periods result from Milankovictch cycles – long periodicities in Earth's orbital parameters.

The human genus, Homo, evolved during the early Pleistocene, about 2 million years ago, and our species, Homo sapiens, came into being about 200,000 years ago. We developed language and complex reasoning around 100,000 years ago.

---

[44] A. Hallam and P.B. Wignall, *Mass Extinctions and Their Aftermath* (Oxford: Oxford University Press, 1997), 59.

[45] Mathez and Smerdon, *Climate Change: The Science of Global Warming*, 206.

[46] Mathez and Smerdon, 209.

The Pleistocene glaciation cycles ended with the Eemian interglacial, 125,000 years ago, which was probably warmer than today.[47] Sea levels were 6 to 9 meters higher than they are now. The Eemian ended when the Last Glacial Period began around 115,000 years ago. Maximum glaciation for that cycle occurred about 22,000 years ago. Ice covered 30% of the global land area. Canada was almost completely covered by ice, as was the northern part of the United States.[48] Sea levels were about 125 meters lower than they are today, because so much water was tied up in ice sheets on land.[49] After the Last Glacial Maximum, warming started again about 18,000 years ago. Much of the ice melted and sea levels rose as the temperatures increased.

The last episode before our current epoch is called the Younger Dryas, which was a short (1,200-year) cold period between 12,900 and 11,700 years ago. The Holocene epoch began 11,700 years ago. It is an interglacial period. Its beginning coincides with humans' invention of agriculture. Its relatively warm temperatures have helped civilization flourish.

It has been proposed that we are entering a new geological epoch, the Anthropocene, which started when human activities started to have significant impacts upon the global environment. One proposal dates the start of the new epoch from the first use of the atomic bomb in 1945.[50] The start date could also be pushed back to the beginning of the industrial revolution, or the invention of agriculture.

One possible reaction to the tale of vast changes in Earth's climate over many millions of years is: "Are we sure that the warming we're seeing how isn't part of a natural cycle like the Pleistocene interglacials?" The answer is "yes, we're sure." One reason is that the rate of temperature rise now is much faster than any previous rise. The fastest known previous rate, from the recovery

---

[47] "Eemian," in *Wikipedia*, https://en.wikipedia.org/wiki/Eemian.

[48] "Last Glacial Period," in *Wikipedia*, https://en.wikipedia.org/wiki/Last_Glacial_Period.

[49] Richard Poore, Richard S. Williams, and Christopher Tracey, "Sea Level and Climate" (USGS, n.d.), https://pubs.usgs.gov/fs/fs2-00/.

[50] Meera Subramanian, "Anthropocene Now: Influential Panel Votes to Recognize Earth's New Epoch," *Nature*, May 21, 2019, https://www.nature.com/articles/d41586-019-01641-5.

after the Last Glacial Maximum, was 0.2° per century, while our current rate of increase is about 0.6° per century.[51] And our current CO_2 concentration is substantially higher than any CO_2 concentration seen on Earth during the previous 800,000 years.[52] We also have hundreds of points of confirmation that the temperature increases are caused by human activities.

Another question we might ask after considering the Earth's long climate history is: "Can't we solve the global heating problem by just waiting for the next ice age?" First off, it's not clear that there will be another ice age, given humans' interference with the climate system. Second, if we weren't interfering, we would expect the next ice age about 50,000 years from now.[53]

Mitigation vs. adaptation

Efforts we make to reduce the causes of climate change, such as the emission of greenhouse gases and deforestation, are mitigation. Efforts we make to reduce the effects of global heating, such as building sea walls to protect a coastal area from sea-level rise, are adaptation. Some actions contribute to both mitigation and adaptation. For example, when we manage forests to reduce the chances of forest fires, we are doing adaptation because we are reducing the likelihood of forest fires, one of the effects of global warming. But we are also mitigating, because forest fires are an important source of GHG emissions.

We want to prioritize mitigation over adaptation in our policy choices, because mitigation solves the root problem, while adaptation just helps compensate for its effects.

Modelling Earth's climate

We know a lot about the Earth's recent and distant climate history, and we know that the planet is heating up due to the anthropogenic release of greenhouse gases and other human causes. We continue to burn fossil fuels

---

[51] Houghton, *Global Warming - The Complete Briefing*, 85.

[52] Mathez and Smerdon, *Climate Change: The Science of Global Warming*, 211.

[53] Houghton, *Global Warming - The Complete Briefing*, 84.

and Earth continues to warm. How can we predict the impacts that will occur as the process continues?

Climate models are computer simulations of the Earth's climate system, and can be used to predict what will happen to the climate under a variety of future scenarios. They are similar to the computer models that predict the weather, but they are broader in scope: the largest of them cover the entire Earth and extend many decades into the future. There are a huge number of climate models, ranging from simple models that focus on one aspect in a small region, to comprehensive global models that simulate the atmosphere, land, and oceans world-wide. The Coupled Model Interconnection Project (CMIP) coordinates the modelling efforts of the most significant models and produces predictions based on a weighted average of the most reliable models.[54] The models are tuned and calibrated in part by running historical simulations of the climate between 1850 and the present.

Most models divide Earth's space into cells – a grid over the surface of the planet, and dozens of levels up and down, into the atmosphere, ocean and land. The simulations calculate the state of each cell in each time quantum based on inputs such as solar radiation and the states of adjacent cells. When the cells are small, running a global simulation requires the largest supercomputers, due to the huge amount of computation. Since the climate system is complex and the Earth is not homogeneous, there are big regional differences. The Arctic, for example, is heating up several times as fast as the planetary average. Models are very useful in helping to predict how global heating will affect a particular area. The models have developed to allow the 2021 WGI IPCC report to predict certain impacts across dozens of regions.[55]

Emissions scenarios

An important input to climate models for the future is the set of economic and political choices we'll make in the crucial upcoming decades. The major decisions will be made by governments, in response to political pressure from their constituents.

---

[54] IPCC WGI, "Climate Change 2021: The Physical Science Basis," 215.
[55] IPCC WGI, 10.

To predict the climate in the future, we need to know what measures we will take in the future to reduce global heating. The most extreme measure would be to ban the burning of fossil fuels and to ban adverse land-use changes now. Because of committed warming, this would probably not suffice to keep the average surface-temperature increase below 1.5°C, and it would wreck the economy. The other extreme is "business as usual," where we continue burning fossil fuels and emitting GHGs as we are now. This scenario would result in much larger long-term temperature increases.

The IPCC has used a number of scenarios, known as Shared Socioeconomic Pathways, or SSPs, in its reports. According to the IPCC, "a scenario is a description of how the future may develop based on a coherent and internally consistent set of assumptions about key drivers including demography, economic process, technological innovation, governance, lifestyles and relationships among those driving forces."[56] The scenarios used in the AR6 Synthesis Report include:

- SSP1-1.9 holds warming to approximately 1.5°C after slight overshoot; requires net-zero GHG emissions by around the middle of this century.
- SSP1-2.6 holds warming to 2.0°C, with net zero emissions in the second half of this century.
- SSP2-4.5 is in line with the upper end of states' Paris-Agreement commitments, resulting in best-estimate warming around 2.7°C by the end of this century. 2.7°C long-term (by 2100) temperature increase.[57]
- SSP3-7.0 is a medium-to-high reference scenario resulting from no climate policy additional to the current policies. 3.6°C long-term increase.
- SSP5-8.5 is a high reference scenario with no additional climate policy, assuming a fossil-fueled socioeconomic pathway. 3.5°-7.6°C long-term average global temperature increase, and a 6.2°C-15.2°C increase likely in the Arctic.[58]

---

[56] IPCC WGI, 227.
[57] IPCC WGI, 14.
[58] IPCC WGI, 572.

The above are current best estimates for the year 2100 based on consolidating the results of several climate models. But temperature increases for the year 2300 are projected to be much higher for the high-emissions scenarios: 8.2°C for SSP3-7.0 and 9.6°C for SSP5-8.5.[59]

Given that most States are not meeting their Paris-Agreement commitments ("NDCs" or "Nationally Determined Contributions), my sense is that we're on a path to a 4°C increase in average global surface temperature. We could, with the right policies in place, be on a path to just a 2°C increase, and the difference between the impacts of 2° and 4° will be huge. It's unlikely that we'll exceed the 4° increase because the impacts of that much warming will be so severe that there will be huge political pressure to reduce emissions. But we will, by then, have already committed our future world to live in a 4° world for at least several dozen generations.

Impacts of climate change

This summary of climate impacts will focus on the impacts of the higher-emission scenarios like SSP3-7.0 and SSP5-8.5, as compared to SSP1-2.6. There is still time to achieve the reductions compatible with SSP1-2.6 if we had the political will to do so.

Heat waves

Kim Stanley Robinson's near-term sci-fi novel about climate change, *The Ministry for the Future*, begins with a fictional account of a heat wave that kills 20 million people in India. The human body can dissipate heat fast enough by sweating to keep body temperature at the required 37°C when the relative humidity is below 20%. But humans can't survive when temperatures and humidity are high enough. A country like India has areas with high temperatures and humidity, and is poor enough that many of the inhabitants can't afford air conditioning, so Robinson's scenario is quite plausible.

---

[59] IPCC WGI, 633.

The following graph[60] shows the combinations of high temperature and high humidity that are deadly:

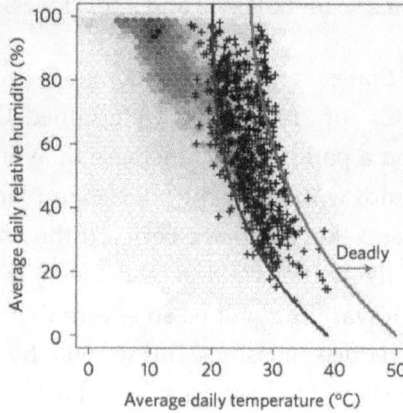

Figure 2-5: *Deadly combinations of heat and humidity*

The black crosses represent deaths. The left curved line is a statistical best fit for the line between lethal and non-lethal events. The rightmost curved line is a statistical best fit for the line denoting a 95% probability of death, so deadly areas of the temperature/humidity graph, where humans are very unlikely to survive, are to the right of the rightmost line. As a rough guide, humans can't survive when the temperature is above 35°C and the humidity is higher than 30%.

The following two graphs show the percentages of land and population, respectively, that will be subject to deadly heat beyond the rightmost line above for more than 20 days per year, under various scenarios. The scenarios used here are the Representative Concentration Pathways (RCPs), which are model trajectories for how GHG concentrations could be reduced over the next century. RPC8.5 corresponds to SSP5-8.5, but the methodology behind them is somewhat different.

---

[60] Camilo Mora et al., "Global Risk of Deadly Heat," *Nature Climate Change* 7, no. July 2017 (June 19, 2017): 502, https://doi.org/10.1038/NCLIMATE3322, used with permission of Springer Nature.

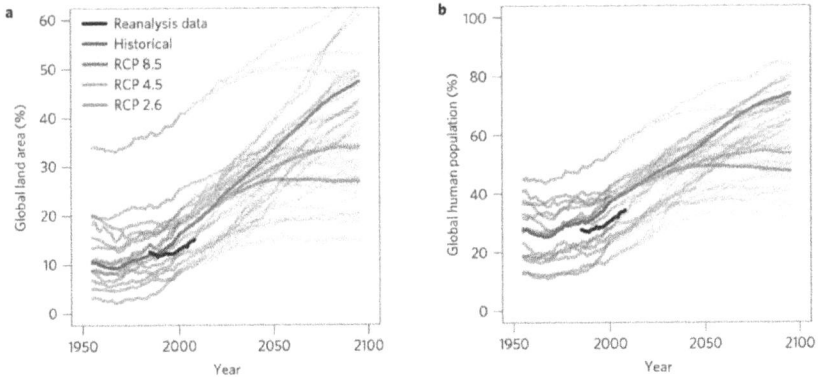

Figure 2-6: *Populations and lands subject to deadly heat*

The left graph shows that, with strong mitigation (RCP 2.6), the portion of our land surface subject to deadly heat more than 20 days per year will be around 25% in 2100, but under RCP 8.5, that portion would be almost 50%. Similarly, the portion of the population subject to deadly heat more than 20 days per year would be around 45% under RCP 2.6, but would rise to 75% under RCP 8.5.[61] Another study estimates the projected increase in the global mortality rate due to climate change at 73 deaths per 100,000 at the end of the century under a high emissions scenario like RCP 8.5. This effect is similar in magnitude to the current global mortality burden of all cancers or all infectious diseases.[62]

Because humidity and temperature tend to be higher in the tropics than in temporal latitudes, the impacts would be felt mostly in the Global South near the equator, as shown on the map below. High emissions like those envisioned in RCP 8.5, mean that the large swaths of land in the tropics, shown in red on the map below, would become uninhabitable, at least without air conditioning, and many of the people who live in the red areas can't afford air conditioning.

---

[61] Mora et al., 502, used with permission of Springer Nature.

[62] Tamma A. et al Carleton, "Valuing the Global Mortality Consequences of Climate Change Accounting for Adaptation Costs and Benefits" (Cambridge, MA: National Bureau of Economic Research, April 2022), 2, https://www.nber.org/system/files/working_papers/w27599/w27599.pdf.

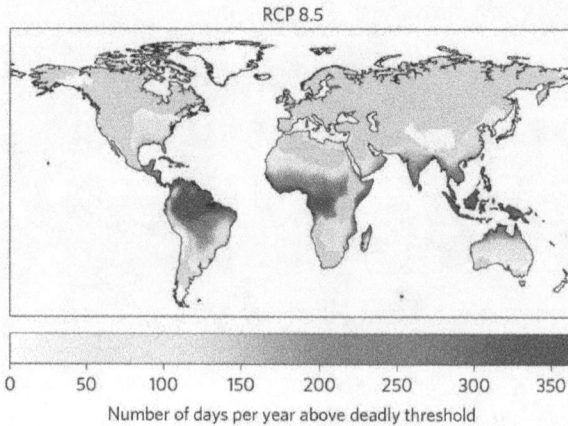

Figure 2-7: *Map of future deadly heat days*[63]

Currently, air conditioning results in a feedback cycle: as temperature increases, more air conditioning is needed, using more electricity. Generating the electricity results in more GHG emissions, increasing temperatures, increasing the need for air conditioning still further. We'll break this feedback loop when all of our electricity is generated from renewable sources.

Increased heat will reduce labor productivity for those who work outdoors. One study estimated that the heat stress from projected 3°C warming above baseline would reduce agricultural labor capacity by 30-50% in Sub-Saharan Africa and southeast Asia, leading to a 5% increase in crop prices.[64]

Increased storm events

Climate Change is increasing the intensity and frequency of extreme weather events, such as heavy rainstorms and hurricanes.[65] As the atmosphere heats up, it absorbs more water vapor, and the extra heat contributes to higher-velocity winds. The proportional increase in rainfall

---

[63] Mora et al., 503, used with permission of Springer Nature.

[64] IPCC WGII, "Climate Change 2022: Impacts, Adaptation and Vulnerability," 725.

[65] Mathez and Smerdon, *Climate Change: The Science of Global Warming*, 284–90.

intensity will be larger than the increase in temperature. Events that just occur now every 10 or 50 years will occur twice or thrice as often, respectively, with 4°C of average warming.[66] We cannot say that specific events were caused directly by climate change, but we can confidently predict that climate change will affect the probabilities of extreme events and make future extreme events even more likely than today's.[67]

Melting ice

Ice is declining worldwide, due to the warming. The volume of sea ice in the Arctic, as measured each September, has declined from 17 thousand cubic kilometers to 4 thousand, between 1979 and 2019.[68] That's a 76% decline. By the end of this century, under the SSP2-4.5 and higher emissions scenarios, the Arctic Ocean will likely be close to ice-free in the summer for the first time in the history of human civilization.[69] The Arctic is a very important ecosystem for the planet and teems with plants and animals adapted to live only there, like polar bears. It's warming about twice as much as the rest of the planet.

Two huge ice sheets, one in Greenland, and one in Antarctica, are likewise melting away. They contain enough volume to raise sea levels 76 meters, if they melted and ran into the ocean.[70] Glaciers worldwide are expected to lose 18% of their mass in this century under RCP2.6, and 36% under RCP8.5.[71]

Sea level rise

Climate change has not only added water to the ocean by melting ice on land, it has heated the ocean, causing the ocean water to expand. Both of

---

[66] IPCC WGI, "Climate Change 2021: The Physical Science Basis," 1567.

[67] Mathez and Smerdon, *Climate Change: The Science of Global Warming*, 287.

[68] Axel Schweiger et al., "Uncertainty in Modeled Arctic Sea Ice Volume," *Journal of Geophysical Research* 116 (September 27, 2011): C00D06, https://doi.org/10.1029/2011JC007084.

[69] IPCC WGI, "Climate Change 2021: The Physical Science Basis," 575.

[70] Mathez and Smerdon, *Climate Change: The Science of Global Warming*, 305.

[71] IPCC, "Special Report on the Ocean and Cryosphere in a Changing Climate," 17.

these factors have led to an increase in global average sea level. The accompanying graph,[72] plotting multiple sets of sea-level calculations as well as a model (the thin solid line in the middle), shows the sea level has risen by about 70 mm (about 2.8 inches) since 1990. This doesn't seem like a lot but, when added to other climate-change impacts such as the increased intensity of storms, is already a big contributor to flooding during storms in coastal areas.

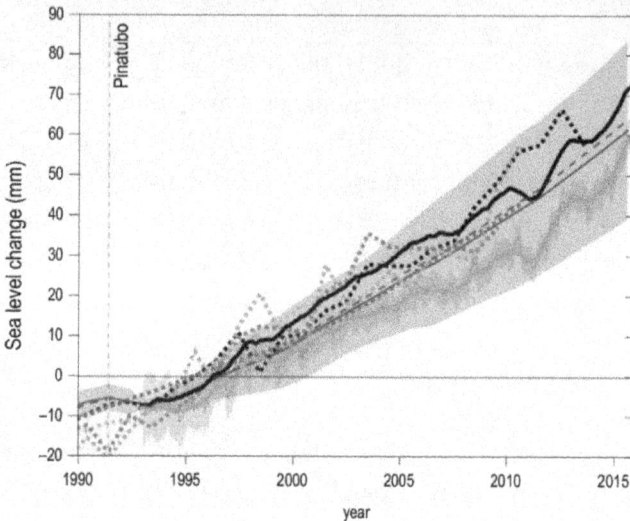

Figure 2-8: *Sea-level rise to date*

Projections of future sea-level rise are based on complex climate models, and there are many uncertainties. Chief among these uncertainties is the likely future of the Greenland and Antarctic ice sheets. The best current

---

[72] Figure 4.5(b) from Oppenheimer, M., B.C. Glavovic , J. Hinkel, R. van de Wal, A.K. Magnan, A. Abd-Elgawad, R. Cai, M. CifuentesJara, R.M. DeConto, T. Ghosh, J. Hay, F. Isla, B. Marzeion, B. Meyssignac, and Z. Sebesvari, 2019: Sea Level Rise and Implications for Low-Lying Islands, Coasts and Communities. In: IPCC Special Report on the Ocean and Cryosphere in a Changing Climate [H.-O. Pörtner, D.C. Roberts, V. Masson-Delmotte, P. Zhai, M. Tignor, E. Poloczanska, K. Mintenbeck, A. Alegría, M. Nicolai, A. Okem, J. Petzold, B. Rama, N.M. Weyer (eds.)]. Cambridge University Press, Cambridge, UK and New York, NY, USA, pp. 321-445. https://doi.org/10.1017/9781009157964.006.

science predicts average sea-level rise of 0.1 to 0.6 meters by 2050, and 0.2 to 1.6 meters by 2100.[73]

Ocean impacts

Oceans have taken up more than 90% of the excess heat in the climate system due to climate change, and have absorbed between 20% and 30% of the total anthropogenic GHG emissions, resulting in acidification of the oceans.[74] The oceans have huge heat capacity, so this excess heat has resulted in an increase smaller than the increase in global atmospheric temperatures — only 0.88°C to date.[75]

There are heatwaves in the oceans, and global heating is a major contributor to them. These heatwaves can kill or alter ocean-based, ecological communities, can result in toxic algal blooms and in decline in fisheries catch and fish farming.[76]

Coral reefs are harmed both by marine heatwaves, and by increasing ocean acidification.[77] Dissolving CO_2 reduces the ocean's pH, making it more acidic, which harms ecological communities, especially coral reefs. The CO_2 dissolved in the ocean also decreases the availability of calcium carbonate, an important component of coral and shellfish skeletons. About 75% of the world's coral reefs are now thought to be at risk due to decreasing ocean pH and other factors.[78]

Floods and droughts

When the atmosphere is 1°C warmer, it can hold 7% more water vapor.[79] Adding energy in the form of heat to the climate system increases temperatures non-uniformly around the world. For example, temperatures

---

[73] IPCC WGI, "Climate Change 2021: The Physical Science Basis," 1296-97.

[74] IPCC, "Special Report on the Ocean and Cryosphere in a Changing Climate," 9.

[75] IPCC WGI, "Climate Change 2021: The Physical Science Basis," 1221.

[76] IPCC WGI, 1227.

[77] IPCC WGI, 1227.

[78] Mathez and Smerdon, *Climate Change: The Science of Global Warming*, 123.

[79] IPCC WGI, "Climate Change 2021: The Physical Science Basis," 1065.

in polar regions are increasing several times more than average. Increased climate-system energy also impacts the way water and water vapor are distributed in very different ways in different places, increasing it in some places, and decreasing it in others.

Terrestrial freshwater represents about 1.8% of all water on Earth, but 97% of that is contained in ice sheets, glaciers, and snow pack, so easily accessible freshwater makes up only 0.054% of Earth's water.[80] It is probably the most important environmental resource, both for humans and for ecosystems. Adequate freshwater is required to grow food, for example. "Currently, around four billion people live under conditions of severe freshwater scarcity for at least one month of the year, with half a billion people in the world facing sever water scarcity all year round."[81] This freshwater scarcity causes food insecurity for many of these people.

Climate change is shifting the patterns of precipitation, increasing the intensity of heavy precipitation events in some places, which, in turn, will increase the intensity of floods caused by those events.[82] But climate change is also causing droughts in other places because it changes the spatial distribution of precipitation. Increased temperature and the evaporation of soil moisture also contribute.[83] Regions like California that depend for their water on snowpack melting through most of the year can be afflicted with drought when increasing temperature causes the snowpack to melt prematurely, or causes precipitation to fall as rain instead of snow, and drain off quickly.

Wildfires

Global heating increases temperatures, and the increase dries out vegetation. By adding energy to the climate system, climate change also increases wind velocities. These two factors result in increased wildfires. Global heating has, for example, resulted in a doubling of burned area in

---

[80] IPCC WGI, 1060.
[81] IPCC WGI, 1060.
[82] IPCC WGI, 1073.
[83] IPCC WGI, 1073.

the western US from 1984 to 2015.[84] Big wildfires, in turn, release a great deal of CO_2, a positive feedback to the climate system.

Tree mortality

"Water stress, leading to plant hydraulic failure, is the principal mechanism of drought-induced tree mortality."[85] Timber cutting, livestock grazing, air pollution, and other non-climate related factors also contribute. Significantly increased tree mortality has been observed in temperate forests in the western US, the African Sahel and North Africa, but also in tropical rainforests like the Brazilian Amazon. The leading cause of tree mortality in the Amazon region is deforestation—clearing forest land to make way for agriculture and livestock grading. This deforestation is an important driver of climate change because it removes forests that provide significant carbon sinks.[86]

Cities

4.2 billion people, just over half the world's population, live in urban areas now, and an additional 2.5 billion people are projected to be living in cities by 2050.[87] More than a billion people will be living in low-lying cities and settlements expected to be at risk from coastal-specific climate hazards by 2050. Those hazards include sea-level rise, increases in tropical cyclone storm surge, and more frequent and intense extreme precipitation.[88] Climate change continues to significantly increase the risk of flooding in cities. It also reduces water security and food security in cities. Cities are drivers of climate change, generating about 70% of worldwide GHG emissions.[89]

It tends to be hotter within cities than without, due to the urban heat island effect, which amplifies global heating. Extreme heat risks resulting from the

---

[84] IPCC WGII, "Climate Change 2022: Impacts, Adaptation and Vulnerability," 247.
[85] IPCC WGII, 248.
[86] IPCC WGII, 249.
[87] IPCC WGII, 909.
[88] IPCC WGII, 925.
[89] IPCC WGII, 989.

combination of the two are expected to affect half the future urban population.[90]

Extensive adaptation will be required to keep cities livable with increased temperatures. Coastal cities will need to figure out how to adapt to rising sea levels, either by abandoning low-lying land or by armoring the coast with dikes like Holland, to keep the sea out. Cities where heat waves will threaten lives must develop strategies to keep their residents cool during times of extreme heat. Planners need to evaluate whether sufficient water may become unavailable in the future, and take measures to augment the supply or to reduce use during water emergencies.

Good urban planning could help considerably with both climate mitigation and adaptation, but urban development plans often pay lip service to climate concerns without actually constraining development to respond to those concerns. Building infill rather than sprawl, ensuring robust and extensive public transportation, colocation of higher residential and job densities, constructing charging infrastructure for electric vehicles, ensuring there is adequate water supply for new construction, and eliminating the use of natural gas in new construction are requirements that could be adopted now, if local politicians had the will to do so. Such planning is especially important for rapidly growing cities in developing countries, where there is an opportunity to build climate-friendly cities from scratch. 2.5 billion people will be added to urban areas between 2018 and 2050, with 90% of this increase taking place in Africa and Asia.[91]

Agriculture and other food production

Global heating generally harms agricultural production, though some of its impacts may be positive. Increased CO_2 in the atmosphere, and higher temperatures, can result in increased yields for some crops, for example. But, overall, increased temperatures on top of increased droughts and floods reduce yields. For example, the combined effects of heat and

---

[90] IPCC WGII, 924.

[91] IPCC WGIII, "Climate Change 2022: Mitigation of Climate Change," 868.

drought are currently decreasing global average yields of maize, soybeans, and wheat by 11.6%, 12.4% and 9.2%, respectively.[92]

41% of major insect pest species will increase their damage further as climate warms, and only 4% will reduce their impacts.[93]

Partly because of its impacts on food production, global heating will lead to increased food insecurity for tens to hundreds of millions of people, particularly low-income populations in developing countries, under high reference scenarios such as SSP5-8.5. Under these scenarios, up to 183 million additional people are projected to become undernourished in low-income countries by 2050.[94]

Migration and displacement

"Current emissions pathways lead to scenarios for the period between 2050 and 2100 in which hundreds of millions of people will be at risk of displacement due to rising sea levels, floods, tropical cyclones, droughts, extreme heat, wildfires and other hazards, with land degradation exacerbating these risks in many regions."[95]

Human health impacts

Warming increases the geographic range where certain diseases such as cholera can flourish. It also increases the ranges of vectors such as mosquitos which carry infectious diseases like malaria.

Studies predict that an additional 250,000 deaths by 2050 will he caused by climate-sensitive diseases and conditions. Under high-emissions scenarios, 6.8 million additional deaths will be caused by global heating by 2100.[96] These additional deaths will be caused primarily by heat, childhood undernutrition, malaria, and diarrheal disease.

---

[92] IPCC WGII, "Climate Change 2022: Impacts, Adaptation and Vulnerability," 728.
[93] IPCC WGII, 219.
[94] IPCC WGII, 2462.
[95] IPCC WGII, 1083.
[96] IPCC WGII, 1090.

Species loss

"Recent research predicts that one-third of all plant and animal species could be extinct by 2070 if climate change continues as it is."[97] Local extinctions and species shifting their range toward higher elevations and toward the poles, are becoming common. But plants and animals may not have a clear path to move to cooler habitat as temperatures increase. They may be blocked by human infrastructure, or by natural barriers such as rivers and mountains. And they may be badly adapted to the habitat they can reach, which may be different from their habitats of origin. The late, great Harvard biologist E.O. Wilson estimated that 30,000 species per year are currently being driven to extinction.[98] "It is estimated that one-third of all reef-building corals, a third of all fresh-water mollusks, a third of sharks and rays, a quarter of all mammals, a fifth of all reptiles, and a sixth of all birds are headed toward extinction."[99]

The species on this planet represent our biological heritage. They are very precious and can't easily be replaced. It can take millions of years to generate new species. One study found that it takes at least 10 million years for life to fully recover after a mass extinction.[100] The loss of biodiversity is the climate impact from which it will be most difficult to recover. It will take only hundreds or thousands of years to purge our atmosphere of excess GHGs and restore our global temperatures to their pre-industrial levels once we stop emitting GHGs. But it will take millions of years for our biodiversity to recover from the Anthropocene mass extinction, which is well underway.

If we grant non-human animals moral standing, so that impacts on them are ethically significant, then the loss of species due to climate change and

---

[97] IPCC WGII, 221.

[98] Niles Eldredge, "A Field Guide to the Sixth Extinction," *New York Times*, 1999, https://archive.nytimes.com/www.nytimes.com/library/magazine/millennium/m6/extinction-eldredge.html.

[99] Elizabeth Kolbert, *The Sixth Extinction* (Henry Holt, 2014), 17.

[100] Science Daily, "Evolution Imposes 'speed Limit' on Recovery after Mass Extinctions," April 8, 2019, https://www.sciencedaily.com/releases/2019/04/190408114252.htm.

environmental degradation does huge moral harm. Even if we just consider species' instrumental value to humans, the loss is enormous. Many of our drugs come from chemical compounds produced by plants, and eliminating those plant species reduces our ability to source drugs this way in the future. The loss of plant and animal species will upend the ecology in many habitats, and have cascading effects on other species, including humans.

Who is affected most by climate change?

Global heating will strongly impact, at least indirectly, all humans and many other types of animals. It will affect not only the current population of 8 billion humans and untold billions of other animals, but also many generations to come. Most of the impacts will be negative.

The impacts are not higher near emissions sources, in contrast to conventional air pollution. A ton of CO_2 emitted in Los Angeles affects those in Africa as much as a ton of CO_2 emitted in Africa, while a ton of particulates emitted in Los Angeles mostly affects those living near the source of emissions. I often hear references to "frontline communities" in the fight against climate change, but that concept works only for conventional pollutants, where pollution sources are often located in poor communities, often communities of color. We're all frontline communities for climate change.

Let's consider a hypothetical situation: a rich woman owns a $5 million house on the coast, and a poor woman squats in a shack next door. The local sea level is rising, encroaching and eroding the coast, and the local regulators won't allow the rich woman to build a sea wall. Without that sea wall, both houses will be undermined and become uninhabitable in ten years. On whom does the impact fall most strongly? The rich woman will lose her $5 million investment in her house. She can't continue using it and won't be able to sell it. She may have to accept the loss, and buy another house somewhere else. The poor woman will bear the cost of moving, but that won't be much. (Of course, she might not have enough money to move.) The economic costs in this case fall squarely on the rich. But the rich woman has many more resources to help her deal with the loss of her home.

Money makes a big difference, both practically, and for reducing stress and emotional hardship in the situation. It is possible that the emotional and practical impacts will be felt more strongly by the poor woman.

This is an example of the general pattern: those with limited resources will find it more difficult to adapt to the changing climate and will personally suffer more from it. Those with greater resources will suffer economic damages from climate impacts, and will need to pay for mitigation and adaptation. As discussed in the Climate Economics chapter, those costs will be significant.

A very small percentage of humans and other animals will not feel direct negative effects of global heating—those living in a cool place that can accept a few degrees of average temperature increase and are not at risk from heat waves, far enough from the shore that they won't be affected by sea-level rise or more-intense hurricanes, in a place where wildfires don't spread, and with a dependable water supply, so there's no risk of drought. The climate crisis won't have much direct effect on these people, but they constitute a small percentage of the global population. Most people will be directly affected.

And there are places where impacts will be felt much more strongly than average, for example in the Arctic. Temperatures are increasing several times faster than the global average there, and the increase is melting the sea ice, erasing habitats for polar bears and some seal species. The melting sea ice is one of the main causes of the Arctic amplification of climate impacts—when the ice melts, the underlying seawater, which is much darker than the ice and consequently absorbs much more light and heat and warms up the Arctic. Indigenous people, whose livelihood depends on hunting and fishing, are affected the most because of the impacts on wildlife.

About a billion people live on the coast, with elevations less than 10 meters above sea level. An average sea-level rise of 2.3 meters is quite possible by 2100,[101] though the rise will more likely be around 1 meter. One to three feet

---

[101] IPCC WGI, "Climate Change 2021: The Physical Science Basis," 1308.

of rise doesn't sound like a lot, but it will subject land where hundreds of millions of people live to permanent inundation or annual flooding. Many of them will become refugees. The increased sea levels also exacerbate the effects of storms, which themselves will be intensified by all the extra energy global heating has put into the Earth's climate system. When storms hit land, they can bring storm surges, pushing more water onto land, greatly increasing storm flooding. An increase in the sea level will also contaminate fresh water in aquifers near the shore, by forcing salt water into them.

The strength of certain impacts will depend mostly on one's location on earth, for example:

- Increased wildfires
- The spread of malaria and other tropical diseases
- Droughts and extreme weather events such as storms and heat waves

Negative emissions: Carbon sequestration, removal, and storage

The IPCC scenarios that avoid more than 2°C of warming require some sort of carbon dioxide removal (**CDR**) by 2050, to compensate for the GHG emissions that we can't completely eliminate because doing so would be too expensive, or would require innovations that have not yet occurred.[102] The recent 2022 Scoping Plan for Achieving Carbon Neutrality, California's main plan for achieving its climate goals, takes a similar approach, stating that CDR will...be necessary to achieve carbon neutrality.[103] This is problematic, because none of the technologies needed to do this have been shown to be cost-effective, or sufficiently scalable. In fact, this is how our failure to adopt policies that will keep temperature increases below 2°C manifests itself—relying on things that are unlikely to happen. According

---

[102] IPCC WGIII, "Climate Change 2022: Mitigation of Climate Change," 1261.

[103] California Air Resources Board, "2022 Scoping Plan for Achieving Carbon Neutrality," November 16, 2022, 92, https://ww2.arb.ca.gov/sites/default/files/2022-12/2022-sp.pdf.

to an editorial in *Nature*, "It's not hard to see why many climate scientists have dismissed the near-impossible scale of required negative emissions as 'magical thinking'. Or why the European Academies' Science Advisory Council said in a report this month: 'Negative emission technologies may have a useful role to play but, on the basis of current information, not at the levels required to compensate for inadequate mitigation measures."[104]

Some of the big oil companies are touting a future where they provide carbon-removal services. A great fix for climate change would be a little box that takes in air and removes the carbon dioxide, converting the carbon to a benign form that may easily be sequestered (like diamonds). It would be nice if it would not require very much energy to do this, and if the equipment were inexpensive and easily scaled. This ideal box is probably impossible, for thermodynamic reasons. And, in practice, no one suggests undoing the combustion process by separating CO_2 into carbon and oxygen, because the energy cost would be too high: the energy released when material is burned and carbon is combined with oxygen to make CO_2 is the amount that's necessary to unburn the carbon and effect the separation.

But there are proposals to separate out the CO_2 and sequester it in stable geological formations underground. There are two types of proposals: for Direct Air Capture (**DAC**) and for Carbon Storage and Sequestration (**CSS**).

DAC is like the diamond-producing box described above, except that, instead of separating the captured CO_2 into carbon and oxygen, the CO_2 the box separates out of the atmosphere would be sequestered in stable geological formations underground. It's technically difficult to separate out CO_2 from the atmosphere, in part because the concentration is so low. 420 parts per million is a historically high concentration, but it still makes up less than one two-thousandth of the atmosphere. The separation requires a lot of energy, and only small pilot projects have been tried. The costs are projected to be high: between $600 and $1,000 USD per ton of CO_2 removed.[105] In short,

---

[104] Editorial, "Why Current Negative-Emissions Strategies Remain 'Magical Thinking,'" *Nature* 554, no. 404 (2018), https://doi.org/10.1038/d41586-018-02184-x.
[105] IPCC WGIII, "Climate Change 2022: Mitigation of Climate Change," 1266.

DAC is an unproven technology that we can't rely on now to help with the problem. "It cannot be assumed that [DAC] will be able to feasibly be scaled up to address a major fraction of current CO_2 emissions."[106]

CCS, by contrast, has been proven to be effective, even though it's expensive. It's essentially the same process as DAC, but is applied to the gas emitted from power plants — mostly coal-fired, but sometimes natural-gas plants. The effluent gas contains a much higher CO_2 concentration than the atmosphere, making it easier to separate out the CO_2. The CO_2 is liquified for injection below-ground after it's separated out. The process has been proven to work at several power plants, but it's expensive enough that it drives the cost of electricity up way beyond the cost of renewable power.[107] Many existing power plants are in locations where the underlying geology won't retain injected CO_2, so the effluent would need to be transported by pipeline to a distant location to be injected into the ground. Hundreds of power plants would need to be retrofitted to reduce global net GHG emissions by a single-digit percentage.[108] For most existing fossil-fueled electric power plants it will be more cost-effective to replace them with renewable generation than to retrofit them with CCS.

Forests and agriculture

Forests and agriculture are important sources of GHG emissions, but they can also, if properly managed, make a substantial contribution to removing GHGs from the atmosphere. The IPCC reports lump them together with other land-use changes, as "AFOLU": agriculture, forestry and other land uses. Between 2010 and 2019, this sector provided an annual sink of 12.5 $GtCO_2$, but was the source of 5.9 $GtCO_2$ of emissions; adding the two together, AFOLU provides a net sink of 6.6 $GtCO_2$/year.[109] These

---

[106] National Research Council, "Climate Intervention: Carbon Dioxide Removal and Reliable Sequestration" (National Academies Press, 2015), 106, http://nap.nationalacademies.org/18805.

[107] Mathez and Smerdon, *Climate Change: The Science of Global Warming*, 381.

[108] Mathez and Smerdon, 385.

[109] IPCC WGIII, "Climate Change 2022: Mitigation of Climate Change," 750.

contributions are significant, given that worldwide annual GHG emissions are currently around 59 GtCO2e/year.

On the forest side, deforestation is a major contributor to global heating. Forests provide carbon sinks, so removing them removes a carbon sink. In addition, when a forest is cut down, the plant material decays, releasing a substantial amount of GHGs. Conversely, reforestation increases carbon sinks, increasing nature's removal of CO_2 from the atmosphere. The science is not well developed on the best way to manage forests to promote carbon sequestration in trees and the soil; that's an important area for future research.

The same is true on the agricultural side—crops and soils can be managed to sequester carbon, but we still need research on how to do this, and we need policies that encourage farmers to manage their farms to sequester carbon. Agriculture accounts for 89% of global methane emissions between 1990 and 2019. Cattle and other ruminants generate methane as part of their digestion; this process, plus rice cultivation, account for most of the growth in atmospheric methane.[110] Moving toward an environmentally friendly plant-based diet would reduce animal agriculture and consequently reduce methane emissions.

Geoengineering

Earth-scale "fixes" for climate change have been suggested. I'll briefly discuss two of them here. Fossil-fuel producers and others who benefit economically from business-as-usual tout these as potential climate fixes. "Why should we disrupt our business and incur the considerable economic pain of mitigation when we may be able in the future to cheaply deploy one of these fixes to solve the problem?" they ask. This is like the enthusiasm for direct air capture discussed above.

Some folks always think that human ingenuity will provide easy solutions for our big problems, so we don't need to take immediate drastic measures. Sometimes they're right. Paul Ehrlich's book, *The Population Bomb*,

---

[110] IPCC WGIII, 765.

published in 1970, predicted that the Earth would run out of resources, especially food, to support its exponentially growing population within a few decades. This never happened because the green revolution, which greatly improved agricultural practices, increased crop yields in developing countries, providing enough food to allow population growth to continue. But we obviously can't count on unknown future technology and human ingenuity to solve all our problems.

The most prominent geoengineering proposal is to inject sulfur dioxide (SO_2) into the atmosphere, to mimic the effects of a volcanic eruption. The increased SO_2 reflects more sun energy back out into space, cooling the atmosphere. In 1991 Mt. Pinatubo in the Philippines erupted, propelling 20 metric tons of SO_2 into the stratosphere, resulting in a global mean surface cooling of 0.4°C.[111] To use this method to fend off global heating, we would need to keep injecting SO_2 into the atmosphere as GHG concentrations continue to increase. And increasing atmospheric SO_2 doesn't lower the concentration of greenhouse gases, so ocean acidification, for example, would continue unabated.

Another proposal is to add iron to the ocean, which would stimulate the growth of phytoplankton, which would consume CO_2 dissolved in the ocean.[112] This in turn would draw CO_2 out of the atmosphere. But adding significant iron to the ocean would greatly change the dynamics of ocean ecosystems, and we don't know enough about how these ecosystems work to accurately predict the consequences.

There are big risks associated with these and other proposed geoengineering schemes. They haven't been tested at scale, and they affect complex atmospheric and ocean systems, so we can't know all the consequences in advance. And who would decide whether or how we should implement the geoengineering? In one of Kim Stanley Robinson's novels, a nation-state that is being heavily affected by global heating decides on its own to inject SO_2 into the atmosphere, and this sort of tampering with the global environment by a rogue nation will be a risk in

[111] Mathez and Smerdon, *Climate Change: The Science of Global Warming*, 178.
[112] Dessler, *Introduction to Modern Climate Change*, 192.

the decades to come. Ideally, any such action would be taken only by international consensus.

A plus four degree world

The IPCC reports offer a variety of scenarios corresponding to possible Earth climate futures. The Shared Socioeconomic Pathways used in AR6 (discussed above) include SSP1-1.9, which would limit warming to approximately 1.5°C after a slight overshoot and SSP1-2.6, which would probably keep warming below 2°C, and assumes we get to net zero emissions by 2050. My opinion is that we're not reducing our fossil-fuel use, and we need to reduce it a lot, and fast, if we're to achieve either of these outcomes. We're on a path that much more resembles SSP3-7.0 so, unless we change out ways, we can expect 4°C of warming before we wake up and stop burning fossil fuels.

What will life on Earth be like in a 4°C world? Mark Lynas' book *Our Final Warning: Six Degrees of Climate Emergency*[113] summarizes peer-reviewed science on the impacts we can expect with levels of warming between 1°C and 6°C. Here is a summary of the impacts Lynas' book says we can expect in a 4°C world:

- Heat waves: half the Earth's land area and nearly three-quarters of the global population will be exposed to deadly heat for more than 20 days per year.
- Droughts: half the world's land surface will become classed as "arid." An additional three billion people will suffer water stress, with a third of the population no longer having access to sufficient fresh water.
- Fire: a large increase in gigantic wildfires.
- Glaciers will all have melted, and much of the snowpack used to store water will be gone as well. In the long run, with 4°C of warming, the Greenland and Antarctic ice sheets will almost completely melt.

---

[113] Mark Lynas, *Our Final Warning: Six Degrees of Climate Emergency* (London: 4th Estate, 2020).

- Floods: heavier rainfall will result in more floods, especially in east Asia, equatorial Africa and South America. What is currently a once-in-a-century flood now will happen as often as once in a decade.
- Hurricanes: increased energy in the climate system will result in more and more-intense hurricanes.
- Crop failures: heat will kill crops in many areas, including the Corn Belt in the US. Drought will harm crops as well. Simultaneous crop failures in multiple food-producing areas will result in global food shortages.
- Species loss: the last time the Earth experienced 4°C temperatures was around 15–40 million years ago; we will lose at least one-sixth of our species. The Brazilian Amazon, a huge area with perhaps the greatest biodiversity on Earth, will be transformed from a tropical rainforest to a dry savannah shrubland. Many ocean species will be killed by marine heatwaves. The mass extinction will be the largest since an asteroid hit the Yucatan peninsula 65 million years ago, resulting in the extinction of the dinosaurs.
- Sea-level rise: if the world's ice melts and the meltwater goes into the ocean, its level will rise by 30–40 meters. This will submerge all coastal megacities and force two billion people to move. We'll probably see one meter of this rise by the year 2100.

We could end up in a world better or worse than the 4°C one, but, unless we take strong action soon, we seem to be heading for 4°C.

The Gap Reports

The UN Environmental Program (UNEP) produces, each year, an Emissions Gap Report, which compares the actions taken by governments over the prior year with the efforts needed to come into compliance with the 2015 Paris Agreement on climate change, which is discussed in more detail in Chapter 5. The 2022 report finds that "there has been very limited progress in reducing the immense emissions gap for 2030, the gap between the emissions reductions promised and the emissions reductions needed to achieve the temperature goal of the Paris Agreement."[114] The gap results

---

[114] "Emissions Gap Report 2022" (UNEP, 2022), xvi,

from a combination of countries' weak commitments under the Paris
Agreement, and their failure to comply with even these weak commitments.

Global fossil fuel production

GtCO₂/yr

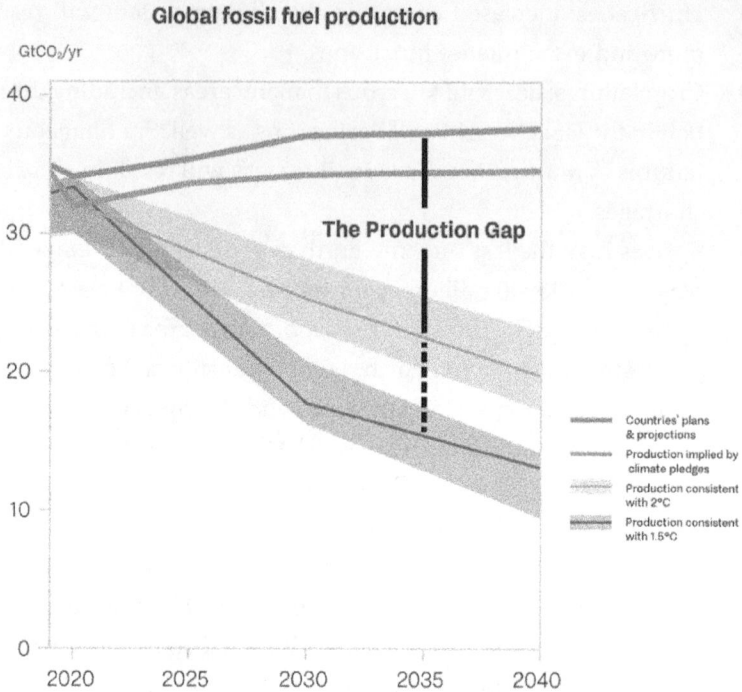

Figure 2-9: *Fossil-fuel production gaps*

The UNEP's Production Gap Report[115] compares countries' projections of
their coal, oil, and natural-gas production with the limitations on burning
fossil fuels that are required, to achieve the Paris-Agreement goals. As you
can see from the Figure 2-9[116], there is a big gap:

This gap shows that countries are not serious about significantly reducing
the burning of fossil fuels. We must plan to reduce the production of fossil
fuels in concert with our reductions in burning them.

---

https://www.unep.org/resources/emissions-gap-report-2022.

[115] UN Environment Programme and Stockholm Environment Institute,
"Production Gap Report 2021" (UNEP, 2021), https://productiongap.org.

[116] Production Gap Report 2021, Figure ES.1, 3.

Big picture future

As long as we continue emitting greenhouse gases, the concentrations of these gases in Earth's atmosphere will continue to increase; global surface temperatures will continue to increase; and the impacts described above will continue to increase.

When we stop emitting GHGs, the level of impact will be determined by the atmospheric GHG concentration at that point. Temperatures will continue to increase, slowly, as the ocean and other systems equilibrate with the atmosphere, but will eventually stabilize. We seem to be on a track toward an increase of 3–4°C, though it would still be possible for us to limit the increase to 2°C if we took immediate, drastic action to reduce our emissions. Impacts at the final impact level will gradually decrease over a period of hundreds to thousands of years, as natural processes remove CO_2 and other GHGs from the atmosphere and the ocean.

Chapter 3
Sustainable Earth

What is sustainability?

A key United Nations document defines "sustainable development" as "development that meets the needs of the present without compromising the ability of future generations to meet their own needs."[1] The main idea behind sustainability is that we humans should be doing only those things that we can continue doing indefinitely, without depleting resources that we or Earth depend on. We are currently using many resources in an unsustainable way. The most important unsustainable action is our release of greenhouse gases (GHGs) into the atmosphere. This release increases the Earth's temperature, which is already too hot. But we're also using too much land, and too much water, allowing topsoil to erode, and polluting the air, soil, and the water. We can't keep doing these things at the current rate.

We can think of the Earth as our home, giving us the duty to maintain it properly. We don't maintain our houses perfectly—there are usually some things that need to be cleaned or repaired—but we don't let them go to wrack and ruin, as we're doing with the Earth. We need to fix a lot of things in addition to GHG concentrations in the atmosphere. There's too much pollution of air and water, too much conversion of natural open space for cities, mines, and infrastructure, too much waste of freshwater, too much topsoil lost to erosion, and too many animal populations being reduced to an unsustainable size. As the United Nations General Assembly recently decreed, all humans have the right to a clean and healthy environment. We need to implement policies that will ensure proper maintenance of our

---

[1] UN World commission on Environment and Development, "Report of the World Commission on Environment and Development: Our Common Future," 1987, 41, https://sustainabledevelopment.un.org/content/documents/5987our-common-future.pdf.

collective home, and it must be done by governments, since there is no economic incentive for private parties to do it.

In addition to depleting resources at an unsustainable rate, we are taking risks at an unsustainable rate. Chapter 2 discussed climate tipping points; if we reach one of these tipping points, we commit ourselves to a large chunk of climate damage in the future. For example, if one of the Antarctic ice sheets is sufficiently loosened by meltwater that it will inexorably slide into the sea over the next few decades, we won't be able to stop it, and we'll have to live with several additional meters of sea-level rise. With each increment of temperature increase we increase the likelihood of reaching these tipping points.

The concept of sustainability can also be applied to risks. For example, if we do something that results in a 1% chance every year of destroying all life on Earth, the likelihood that destruction will come within 100 years is 63%. This is unsustainable.

Long-term thinking is the essence of sustainability. We humans tend to focus on what's happening this week, maybe this year, perhaps a few years out. But human psychology deters us from thinking in terms of centuries, or longer, in the future. I think that a good viewpoint for sustainability thinking is a thousand years from now, the year 3000. But we could sensibly expect humankind to last much longer into the future, as will be discussed below.

Before we talk about the distant future, we need to look at the distant past, to see how we got where we are.

History of humanity in three stages

Three major steps have led us to where we are today:

The first step was life arising on this planet, which happened around 3.5 billion years ago. The first life was a collection of molecules that metabolized energy and reproduced itself. It was important that DNA, which is a template for reproducing a new instance of the organism, could be altered by errors in copying or external events, introducing random variations. This is the basis of evolution. Biologists counsel us not to see evolution in teleological terms: it doesn't work toward evolving more complex organisms. Instead, it moves in all directions at once, as each branch or tendril of the tree of life tries to grow into an ecological niche where it can thrive. Our human species, Homo sapiens, evolved around 300,000 years ago, roughly in the last 0.01% of the time there has been life on Earth.

The second major step was the agricultural revolution, about 12,500 years ago. Before the agricultural revolution, humans were hunter-gatherers, operating in small nomadic bands. We were constrained to our ecological niche like other animals. When we developed agriculture, we escaped our niche. Growing food, instead of finding food, allowed us to greatly increase our population. It gave us the power to alter the Earth in important ways. We've cultivated a lot of land, and the cultivated land is no longer natural habitat for other animals.

The third major step was the scientific and industrial revolution. The scientific revolution started around the year 1500, and was a shift from deductive reasoning from first principles (mainly religious) to our current scientific method. Until the scientific revolution arrived in Europe, China was ahead of the West in science and technology—they had invented printing with movable type, paper, and gunpowder, for example. But as a result of the scientific revolution, Europe pulled ahead, and its science led to the development of technology that fueled industry.

The industrial revolution built upon new scientific discoveries and transformed our infrastructure. James Watt's development of a viable commercial steam engine in the late 1700s is often cited as its start. The industrial revolution resulted in new industrial manufacturing processes, substituting mechanical processes, initially driven by water power, then by steam engines, for hand production methods.[2] It continued with mechanization of transport, initially via railroads, then with automobiles.

---

[2] "Industrial Revolution," in *Wikipedia*, September 6, 2021, https://en.wikipedia.org/w/index.php?title=Industrial_Revolution&oldid=1042697673.

The industrial revolution was fueled with fossil fuels—coal, oil, and natural gas. This energy is distributed in fungible, concentrated form: by the electric grid, natural-gas pipelines, and coal, oil, and gasoline shipped by rail and truck.

Just as agriculture allowed humans to escape their hunter-gatherer ecological niche, the ability to harness mechanical energy allowed us to expand our industrial production beyond the craft production of homes and small businesses. Development of electronic devices, including computers, is a continuation of this process.

Future timescales

How long will Life on Earth and the human race last? The average life of a mammalian species is about a million years.[3] But humans won't go extinct the same way that other animals do—we have technology to shield ourselves from ecology. Species are going extinct at a high rate now because the changing climate, and environmental pollution, are making their habitats unsuitable for them, but we, uniquely, can modify our habitat. We have buildings and other infrastructure that effectively makes new habitat for us anywhere on Earth.

How many humans will live in the future? About 140 million babies are born every year, globally. If this continued for a million years, there would be 140 trillion humans born in the future, 17,500 for every person alive today. The number of future humans could actually be much larger than this—Earth will remain habitable for hundreds of millions of years. But of course the number could be much smaller, if some disaster befalls us.

Existential risks

How do we estimate the probability that humans will be alive and flourishing a million years from now? We humans would be hard to kill.

---

[3] ScienceDirect, "Life Span," *ScienceDirect* (blog), 2013, https://www.sciencedirect.com/topics/earth-and-planetary-sciences/life-span.

We could easily continue even if 99% of the population was killed, leaving a population of only 80 million. But some catastrophes could kill us all, or could damage us or the Earth ecosystems and resources we depend on so much that we would never recover as a civilization.

There are several ways we could cause our own extinction or permanent decline much sooner than a million years from now. As discussed in Chapter 2, it is very unlikely that climate change will cause our extinction, or even a civilizational collapse. But there are other big risks.

One major risk is large-scale nuclear war. Direct impacts—destruction of cities by nuclear explosion—wouldn't kill a major portion of humans; the main risk is from the so-called nuclear winter. Bomb explosions could loft enough dust and soot into the stratosphere to cool the planet up to 7°C for five years,[4] greatly reducing agricultural production. It would be catastrophic, but very unlikely to result in human extinction.[5] The risk of catastrophic nuclear war has declined since the end of the cold war, but, from our long-term sustainability perspective, that's a short reprieve, and the risk will keep growing. Proliferation is a one-way process that can't be stopped in the long run. More countries will get nuclear weapons, and the geopolitical alliances will keep changing. The recent Russian invasion of Ukraine has highlighted the risk that a nuclear-armed country will decide to use nuclear weapons—either tactically on the battlefield or strategically against an opponent's cities—if it's losing a conventional war. In any case, Martin Hellman, a Stanford professor emeritus, estimates the risk of full-scale nuclear war at around 1% per year.[6] This compounds to a rate of 63% per century, far too high. An all-out nuclear war would probably solve the climate crisis, but in a bad way.

---

[4] Toby Ord, *The Precipice: Existential Risk and the Future of Humanity* (New York: Hachette, 2020), 98–100.

[5] Ord, 99.

[6] Martin E. Hellman and Vinton G. Cerf, "An Existential Discussion: What Is the Probability of Nuclear War?," *Bulletin of the Atomic Scientists*, March 18, 2021, https://thebulletin.org/2021/03/an-existential-discussion-what-is-the-probability-of-nuclear-war/#post-heading.

Another major risk comes from pathogens, either natural or engineered. We're still living through the impacts of the natural COVID-19 pandemic, caused by a virus that moved from bats to humans. A bigger, existential, risk comes from man-made pathogens. Fifteen countries maintained bioweapons programs during the twentieth century, including the US, U.K, France, and Russia.[7] In spite of supposedly stringent precautions, there were a number of incidents where dangerous pathogens escaped into the wild from laboratory environments.[8] DNA technology has improved and become much cheaper, so that genetic manipulation is now available to amateurs. Al Qaeda, for example, developed a bioweapons program.[9] Experts estimate the probability of an extinction-level pandemic at between one and three percent per year,[10] a very high—and unsustainable—level. This is probably our largest source of extinction risk.

Bill McKibben,[11] William MacAskill,[12] and Toby Ord[13] all count strong, artificial general intelligence (AGI) as a significant threat to humans. By AGI, we mean the ability of a computer-based system to match and exceed every aspect of human intelligence, so the AGI system can perform any task humans can perform as well or better than humans.[14] They point to the exponential increase in computing capacity that has obtained for several decades, and the ruminations of luminaries such as Ray Kurzweil on transferring human minds into silicon-based computers. They worry that AGI systems unaligned with the goals for which they were created may take over the world and humans will become their domestic animals. I'm very skeptical about the early timetable put forth by some artificial-intelligence (AI) experts: they think AGI may be developed as early as ten

---

[7] Ord, *The Precipice: Existential Risk and the Future of Humanity*, 132.

[8] Ord, 131.

[9] William MacAskill, *What We Owe the Future* (New York: Basic Books, 2022), 131.

[10] MacAskill, 113.

[11] Bill McKibben, *Falter: Has the Human Game Begun to Play Itself Out?* (New York: Holt Paperbacks, 2019), 154–62.

[12] MacAskill, *What We Owe the Future*, 86–91.

[13] Ord, *The Precipice: Existential Risk and the Future of Humanity*, 140–41.

[14] Ord, 140–41.

years from now.[15] And I'm also skeptical about achieving AGI even in the long term. Computers can't exactly simulate even the simplest physical systems, like the atom, let alone the most complicated physical systems, like the human brain. Maybe there's some physical-brain aspect of human intelligence that can never be transferred to electronic computers, so there will always be some areas in which human thinking is more competent.

In addition to existential risks, which threaten the existence of the human species, we should consider risks which threaten us with civilizational collapse, and perhaps the consequent risk that civilization may not be able to recover. The existential risks discussed above, if they occurred at a level not quite severe enough to cause our extinction, nonetheless could cause our civilization to collapse.[16] The production of high-tech devices, like cell phones, that require many components produced with difficult-to-obtain materials, such as rare earths, or that require advanced manufacturing, such as high-density electronic circuits, would be difficult to restart.

We need to lower the level of our existential risks, to make them sustainable, i.e. small enough that we can assume the risks. The risks of large-scale nuclear war and of engineered pandemics are high enough to be unsustainable in the long run. We should take action to reduce these risks now. And we need to keep an eye on the development of AGI, to make sure it doesn't risk getting out of control.

The moral weight of future humans

How much moral weight do we give future humans, compared to the moral weight of humans living right now? It's likely that there will be many more humans in the future than have lived to date, or are living right now. This is similar to the economic issue of discounting, which will be discussed in the next chapter. If we give full weight to future humans, by treating a prospective human life two centuries from now as the same as the life of a person now living, then the moral weight of the future becomes essentially infinite, 17,500 times as much as the weight of current humans.

---

[15] MacAskill, *What We Owe the Future*, 90.
[16] MacAskill, 121–42.

But climate change won't last a million years, so the number of future humans it affects will be much smaller. Let's assume we gradually ramp down GHG emissions to zero by 2050, and by then we have increased atmospheric GHG concentrations so that the global surface temperature has increased by 3°C (a likely scenario). Once we stop emitting GHGs, it will take about 10,000 years for concentrations to be reduced by half; they will be reduced to just a few percent in 100,000 years.[17] So we can use 50,000 years as a rough estimate for the period of time it will take for the Earth to return to normal once we stop emitting GHGs. This means that "only" 7 trillion (140 million × 50,000) future people will be affected. This is 875 times as many people as are living now.

How do we decide on the proper balance between the interests of current people and future people? The previous paragraph shows that many more future people than current people will be affected by climate change.

One attempt to deal with this problem is based on the "non-identity problem," which is a philosophical argument that future people don't have identities, and therefore can't be harmed.[18] The argument goes like this: a million factors affect which sperm will fertilize which egg in the future, and each combination will make a different person, so the identities of persons are in flux until conception. We can't harm a person who has no identity; to harm is to harm an actual person. This seems like a silly argument to make in the context of climate change, at least. If we know that something will harm a hundred people in the future, why does it matter that we don't know exactly who those people will be? How could we believe that, essentially, people who have not been born yet cannot be harmed by our actions?

Even if we reject—as we should—the idea that future humans, lacking identity, should not figure into our climate decisions, we still have to deal with the essentially infinite number of future humans. How do we balance their interests against the current population's?

---

[17] N.S. Lord et al., "An Impulse Response Function for the 'Long Tail' of Excess Atmospheric CO2 in an Earth System Model," *Global Biogeochem Cycles* 30 (January 9, 2016): 8, https://doi.org/10.1002/2014GB005074.

[18] David Boonin, *The Non-Identity Problem & the Ethics of Future People* (Oxford: Oxford University Press, 2014).

I think the answer lies in the combination of uncertainty about the future and contingency. How well can we imagine the lives of humans living in another thousand, ten thousand, or a hundred thousand years? People living just 500 years ago would have been completely unable to imagine the lives of those of us living well in developed countries today, with all of our technology. As argued in the economics chapter of this book, technological progress will slow down in the future because its current growth is based on exponential growth in research effort, which is inherently unsustainable. The difference between now and the future 500 years from now is unlikely to be as great as the difference between 500 years ago and now, but it's still likely to be different enough that we don't have a clear picture of what it will be like. Maybe there will be technological advances that help deal with the climate impacts we're causing.

Second, the future is at least somewhat contingent. What we do about climate now will matter little if humans are wiped out by a pandemic in the next hundred years. The quick tour of existential risks just above suggests that there is a substantial continuing risk—a few percent per year—of human extinction or civilizational collapse. This risk reduces the expected value, in a statistical sense, of the number of future humans. Factoring this risk into the equation, in a rough way, I estimate the moral weight of future humans to be equal to that of humans currently living.

Exponential growth is not sustainable

A quantity grows linearly with time if a constant amount is added to it in each time interval. A quantity grows exponentially if it multiplied by a constant amount in each time interval. The following graph illustrates the difference: the quantity 4 is added to the grey line in each interval to illustrate linear growth: 4, 8, 12, 16, 20, 24. The black-line value is multiplied by 2 in each interval to illustrate exponential growth: 2, 4, 8, 16, 32, 64.

The two lines diverge widely for higher values. At the 6th interval shown above, the linear value is 24, and the exponential value is 64. At the 20th interval, the linear value would be 80, the exponential value would be

1,048,576. At the 100th interval, the linear value would be 400, and the exponential value would be around 10^{30}.

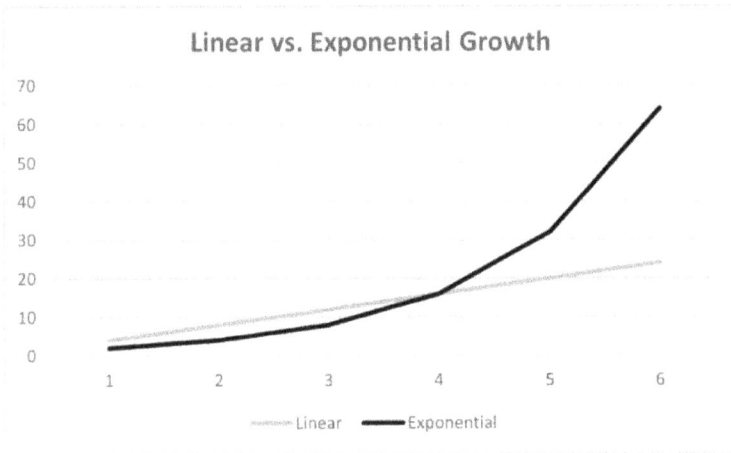

Figure 3-1: *Linear vs. exponential growth*

We are in a period of rapid expansion, innovation, and growth, and have come to consider exponential growth as the norm. Two examples are GDP growth and Moore's law. As discussed at more length in the following chapter, we have come to expect Gross Domestic Product (GDP), the economic value of goods and services produced, to increase by a few percent each year. And Moore's law says that the number of transistors in a dense integrated circuit—and the resulting computing power of the IC—will double every two years.[19]

But exponential growth can occur only in bursts in the short run; it cannot continue in the long run. We humans are not used to considering the long run, but that is, in large part, what this book is about. It uses the year 3000 as a point from which to look back, but that's only a thousand years in the future. We expect to continue our human civilization for tens or hundreds of thousands of years, maybe millions.

If our GDP were to continue growing at 2% per year for the next thousand years, GDP would be about 400 million times today's GDP a thousand

---

[19] Chris Mack, "The Multiple Lives of Moore's Law," *IEEE Spectrum*, March 30, 2015, https://spectrum.ieee.org/the-multiple-lives-of-moores-lawwww.mooreslaw.org.

years from now. We have grown accustomed to this level of GDP growth, to the point where we consider out economy ailing if we don't achieve it. But it's not sustainable.

As for Moore's law, a doubling every two years amounts to a 41% annual growth rate. This is huge, but strong-AI proponents, such as Ray Kurzweil, expect it to continue long-term. By 2099, "a penny's worth of computing power will be a billion times as powerful as all the human brains now on the planet."[20] A 41% growth rate compounded for a hundred years results in an increase by a factor of a quadrillion. I'm very skeptical about Moore's law continuing long-term, even when it is re-expressed in terms of computing power per dollar instead of transistor density on integrated circuits. We may be approaching the physical limits on transistor density, and, as discussed in the Economics chapter of this book, the research effort to support the continuation of Moore's law is itself unsustainable because it's been increasing exponentially in recent years.

Population × consumption = GDP

It has been an environmental truism for decades that environmental damage is proportional to the product of population and consumption. But different types of consumption have different impacts. Burning fossil fuels and eating meat are types of consumption that are harmful to the environment. But gardening, for example, though it involves economic consumption—purchases of seeds, plants, fertilizer, tools, and water— more than offsets the negative environmental effects of producing these products by the environmental benefits it provides as habitat, and in carbon sequestration. And paying lawyers like me to fight climate change via litigation is economic consumption, but the result of a successful lawsuit is a benefit for the environment.

The product of population and per-capita consumption, which environmentalists correlate with environmental harm, is also the mathematical formula for gross domestic product (GDP), which measures total consumption of an economy. This leads some environmentalists to call

---

[20] McKibben, *Falter: Has the Human Game Begun to Play Itself Out?*, 136.

for an end to GDP growth; but we'll want to take a more nuanced look at both the population and consumption components.

Population

Is it better to have a larger or a smaller global population? If we were able to agree on an optimal population for the Earth, we could use non-coercive means such as education, economics, and birth control to achieve it. But we're far from agreeing; some think our population should be much larger, some think it should be smaller.

Ethics: The value of a person

Is it a positive good to bring a new, happy person into the world? Here, I'm talking about a general moral benefit, not the value of the new person for parents and others. The issue is not even whether the world as a whole is better off with the addition of a new, happy person. The issue is whether there is a moral value to having one more life, without considering the effects of the new life on others.

The quality of the added life is important. Adding a person who would be tortured horribly his whole life would not be a moral good, but we're assuming that the added person lives in circumstances where he or she can be reasonably happy and enjoy life.

There has been a lot of discussion among philosophers on this question. Some of them had an intuition that bringing a new person with a good life into existence is morally neutral—neither good nor bad, the so-called "intuition of neutrality."[21] The Oxford philosopher John Broome wrote a book on this topic, trying to find arguments in support of this intuition, but ultimately decided the intuition must be wrong.[22] The British philosopher Derek Parfit, in his book *Reasons and Persons*, discussed this question at some length,[23] but also failed to come to a definitive conclusion.

---

[21] MacAskill, *What We Owe the Future*, 171.

[22] John Broome, *Weighing Lives* (Oxford: Oxford University Press, 2004), preface.

[23] Derek Parfit, *Reasons and Persons* (Oxford: Oxford University Press, 1984), 381–438.

Many of the arguments on this topic are pseudo-quantitative analyses of well-being or happiness summed across different types of hypothetical populations. They involve questions like "would the world better off with a larger population that is, on average, less happy?" and analyze it by multiplying the hypothetical populations by the corresponding well-being numbers. These are a lot like economic arguments, but at least the economists have a concrete, specific measure of utility, namely income. It is proper to quantify income, though economists tend to assume too much that it represents happiness or well-being. But I don't think it's proper to quantify well-being, because one person's well-being may be qualitatively quite different from another person's. Well-being is complicated and multi-dimensional, and not subject to being summed up across groups.

My answer to this question of whether adding a new, high-quality life is a positive moral good derives from life being the basis for all values, ethics, and morality. Without life, we would have a cold, mechanical universe, devoid of value. If our values of what is good and what is not good are based on life, then surely adding another life to the universe must be good, because it increases life.

Optimal global population

To decide on an optimal population for the Earth, we need to balance the factors.

In favor of increasing population:

- Adding more lives is a positive moral good, as discussed just above.
- Increased population means more innovation, which will contribute to economic growth.[24]
- Economic growth is good for the economy and improves people's lives.

---

[24] MacAskill, *What We Owe the Future*, 152–53.

In favor of decreasing population:

- Population makes demands made on limited natural resources.
- For some, lack of resources will lead to reduced quality of life.

To find an optimal population level, we need to find the right balance between these two sets of factors. I would favor only non-coercive means, such as education and access to contraception, to adjust population levels. Such a program will be discussed in the last chapter of this book. In this chapter, we're considering how to decide on an optimal population for the planet. Reviewing the factors just above, the answer seems to be: we should have the largest population that is sustainable, the largest we can have without depleting natural resources, or bringing people into being under circumstances that will lead to their having lives not worth living.

In 1798, Thomas Robert Malthus published a book titled *An Essay on the Principle of Population*, in which he famously claimed that population increases exponentially, but food production increases arithmetically, so that population growth will eventually outstrip food supply, leading to massive famine. This argument was updated in a 1968 book titled *The Population Bomb* by Stanford University Professors Paul and Anne Ehrlich. But of course these predictions have not come to pass, largely due to innovations in agriculture that greatly increased farm productivity.

And it looks like we're on a path toward limited population growth by the end of this century. The current global population is just over 8 billion, and the United Nations estimates that the human population will grow to 8.5 billion in 2030, 9.7 billion in 2050, peaking at 10.4 billion in the 2080s, and remaining at that level until 2100.[25] Most of the growth will occur in developing countries in Asia and Africa; fertility rates in developed countries are lower than the replacement rate, so the populations of

---

[25] United Nations Dept. of Economic and Social Affairs, Population Division, "World Population Prospects 2022: Summary of Results," 2022, i, https://www.un.org/development/desa/pd/sites/www.un.org.development.desa.pd/files/wpp2022_summary_of_results.pdf.

developed countries are expected to decline, unless immigration makes up the difference.[26]

Joel E. Cohen's 1995 book, *How Many People can the Earth Support?*,[27] surveys estimates of the maximum population that Earth can support, starting with a 1679 estimate by Antoni van Leeuwenhoek, the Dutch inventor of the microscope. He estimated Earth's maximum population at 13.4 billion people.[28] Subsequent estimates have ranged from around 1 billion to a trillion, with a median estimate around 10 billion.[29] Cohen refers to the theoretical maximum population as the Earth's "carrying capacity," an ecological term which is defined as "the maximum number of individuals of a given species that the resources available in a given environment can sustainably support."[30] The carrying capacity for humans would be much smaller if we had remained hunter-gatherers; we would be top predators, like lions, and each of us would need a large territory. Our agriculture and technology allow us to live at much higher densities, and the limitations now come not from how much prey we can find within our territory but how many of Earth's resources we can draw upon without depletion.

Even after we've looked at depletable resources (next), we still won't be able to come up with a single number for the optimal population of Earth. For one thing, it depends on policy choices we make or could make, and technological innovations that make more efficient use of resources. And optimal population varies across countries, cultures, and geographies. Adding a person in Denmark is very different from adding a person in Sierra Leone. It seems likely that Earth could sustainably accommodate a population of 10.4 billion persons if we eliminate some of our unsustainable practices, and manage to convince developing countries to adopt sustainable environmental programs as they grow.

---

[26] United Nations Dept. of Economic and Social Affairs, Population Division, 14.

[27] Joel E. Cohen, *How Many People Can the Earth Support?* (New York: W. W. Norton & Company, Inc., 1995).

[28] Cohen, 16.

[29] Cohen, 212–16.

[30] "Carrying Capacity," in *American Heritage Dictionary* (Houghton Mifflin Harcourt, 2020).

Unsustainable resource depletion

We are depleting some of Earth's resources at a rate, in many cases, higher than the replenishment rate. This is unsustainable—we'll eventually run out of these resources.

Water: Only about 0.5% of Earth's water is available as freshwater for human use. This water becomes more polluted every year. More than a billion people still do not have access to fresh water, and more than 2 billion people live in countries experiencing high water stress.[31] Aquifers around the world are being depleted at an unsustainable rate.[32]

Food: Land degradation, declining soil fertility, unsustainable water use, desertification, disease, pest, overfishing and the marine environment degradation will continue to lessen the ability of the natural resource base to supply food.[33]

Species Loss: As can be seen from Figure 3-2, taken from the IPCC WGII Report,[34] more than 50% of species will go locally extinct in areas of South America, southern Africa, Australia, Russia, and the Arctic if the global temperature increase reaches 3°C or 4°C. An increase in the extinction rate beyond the pre-industrial speciation rate (the rate at which new species are created) is unsustainable—the Earth will gradually, over time, lose biodiversity. The background extinction rate has been credibly estimated at one tenth of an extinction per million species per year. "Current extinction

---

[31] United Nations, "Goal 12: Ensure Sustainable Consumption and Production Patterns," Sustainable Development Goals (United Nations), https://www.un.org/sustainabledevelopment/sustainable-consumption-production/.

[32] Margaret Robertson, *Sustainability: Principles and Practice*, 3d. (New York: Routledge, 2021), 119–21.

[33] United Nations, "Goal 12: Ensure Sustainable Consumption and Production Patterns." https://www.un.org/sustainabledevelopment/sustainable-consumption-production/

[34] IPCC WGII, "Climate Change 2022: Impacts, Adaptation and Vulnerability" (IPCC, 2022), 258.

Projected loss of terrestrial and freshwater biodiversity
compared to pre-industrial period

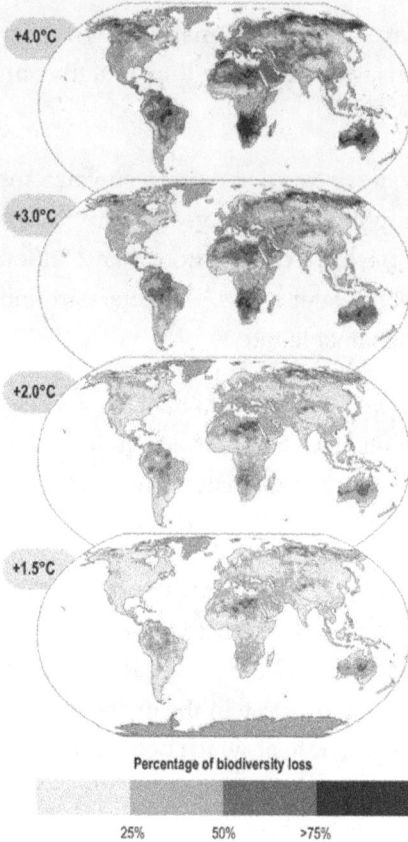

Percentage of biodiversity loss

25%   50%   >75%

rates are 1,000 times higher than natural background rates of extinction and future rates are likely to be 10,000 times higher."[35]

Air Pollution: The World Health Organization estimates that around 7 million people die every year from exposure to fine particulates in polluted air.[36] And around 90% of people worldwide breath polluted air. Actions we take to fight climate change will have large co-benefits in reducing conventional air pollution. Switching from fossil fuel-based power plants to renewable generation, and from petroleum-powered cars to electric cars will greatly reduce pollution.

Figure 3-2: *Biodiversity loss for different areas at increasing levels of climate change*[37]

---

[35] Jurriaan M. De Vos and et al., "Estimating the Normal Background Rate of Species Extinction," *Conservation Biology* 29, no. 2 (2014): 452–62.

[36] World Health Organization, "9 out of 10 People Worldwide Breathe Polluted Air, but More Countries Are Taking Action" (World Health Organization, May 2, 2018), https://www.who.int/news-room/detail/02-05-2018-9-out-of-10-people-worldwide-breathe-polluted-air-but-more-countries-are-taking-action.

[37] Figure 2.6 from Parmesan, C., M.D. Morecroft, Y. Trisurat, R. Adrian, G.Z. Anshari, A. Arneth, Q. Gao, P. Gonzalez, R. Harris, J. Price, N. Stevens, and G.H. Talukdar, 2022: Terrestrial and Freshwater Ecosystems and their Services. In: *Climate Change 2022: Impacts, Adaptation, and Vulnerability*. Contribution of Working Group II to the Sixth Assessment Report of the Intergovernmental Panel on Climate Change [H.-O. Pörtner, D.C. Roberts, M. Tignor, E.S. Poloczanska, K. Mintenbeck, A. Alegría, M. Craig, S. Langsdorf, S. Löschke, V. Möller, A. Okem, B. Rama (eds.)].

Land Degradation and Desertification: A lot of our agricultural land is being harmed, largely by erosion of topsoil. "About a quarter of the Earth's ice-free land area is subject to human-induced degradation.... Soil degradation from agricultural fields is estimated to be currently 10 to 20 times (no tillage) to 100 times (conventional tillage) higher than the soil formation rate."[38] In dryland areas, desertification reduces agricultural productivity and contributes to a loss of biodiversity.[39]

Animal Agriculture: Diets in developed countries contain a lot of meat, and producing meat requires far more water and food inputs than producing equivalent vegetarian calories. Producing a pound of beef, for example, requires ten pounds of cattle feed. About 40% of grain worldwide is used to feed animals, and this grain could be used to directly feed humans. If the population of developing countries adopts the heavy-meat diet of developed countries, the climate impacts will be severe, and we are likely to have a food shortage.

Deforestation: The Earth loses about 10 million hectares of forest every year mostly to agricultural expansion.[40] We need the forest for habitat, but also because it provides a carbon sink that absorbs and sequesters carbon dioxide.

Waste: "The world generates about 2.01 billion tonnes of municipal solid waste annually, with at least 33 percent of that—extremely conservatively—not managed in an environmentally safe manner."[41] 37% of waste is disposed of in a landfill. Placing this much waste into landfills is not sustainable. Conventional landfills generate significant methane

---

Cambridge University Press, Cambridge, UK and New York, NY, USA, pp. 197-377, doi:10.1017/9781009325844.004.

[38] IPCC, "Climate Change and Land" (IPCC, August 7, 2019), 3, https://www.ipcc.ch/srccl/.

[39] IPCC, 3–3.

[40] UN Food and Agriculture Organization, "The State of the World's Forests" (UN FAO, 2020), xvi, https://www.fao.org/3/ca8642en/ca8642en.pdf.

[41] The World Bank, "Trends in Solid Waste Management" (World Bank), https://datatopics.worldbank.org/what-a-waste/trends_in_solid_waste_management.html.

emissions, due to decomposition of organic matter in them. This organic matter should be diverted and composted.

Materials: Some materials, like wood, are renewable, and, to be sustainable, must not be consumed, over the long run, at a rate greater than their replenishment rate. Some materials, like metal ore, are nonrenewable because they cannot be replenished. For many types of nonrenewable materials, the effort and resource requirements to obtain them increase as the supply dwindles. For metals, in many cases, the more concentrated ores have been mined, so a larger quantity of less concentrated ores needs to be removed and refined, taking more effort and energy, and generating much more rock waste.[42] A similar thing is happening with rare earth metals, 97% of which are mined in China. They're important for producing flat-screen monitors, cell phones, batteries, and electric vehicles, and they are being rapidly exhausted. In 100 out of 144 countries, we're consuming materials at an unsustainable rate.[43]

A 2018 article in *Nature Sustainability* analyzed which countries meet 12 social thresholds, such as nutrition, sanitation, and life satisfaction, and which countries had passed biophysical boundaries, such as material footprint, ecological footprint, and CO_2 emissions. It found that countries that met most of the social thresholds, and therefore had a high level of wellbeing, exceeded most of the biophysical boundaries.[44] In other words, the lifestyles of the countries providing the best lives are not sustainable. It also found that basic needs could be met universally in a sustainable manner: "Physical needs (that is, nutrition, sanitation, access to energy and elimination of poverty below the USD\$ 1.90 line) could likely be met for 7 billion people at a level of resource use that does not significantly transgress planetary boundaries. However, if thresholds for the more qualitative goals (that is, life satisfaction, healthy life expectancy, secondary education, democratic quality, social support and equality) are to be universally met then provisioning systems—

---

[42] Robertson, *Sustainability: Principles and Practice*, 341.

[43] Daniel W. O' Neill and et al., "A Good Life for All within Planetary Boundaries," *Nature Sustainability* 1 (February 2018): 90.

[44] O' Neill and et al., "A Good Life for All within Planetary Boundaries."

which mediate the relationship between resource use and social outcomes—must become two to six times more efficient."[45]

Consumption is not all bad

As mentioned above, population × consumption is a very rough proxy for environmental resource depletion, advocated by many environmentalists. By "consumption," we mean per-capita consumption in this equation, so population × consumption is equivalent to GDP. But not all consumption depletes environmental resources. Burning $5 worth of gasoline by driving 30 miles harms the environment much more than reading a $5 book on a Kindle e-reader.

Instead of trying to reduce per-capita consumption per se, we should focus on reducing our unsustainable use of environmental resources. To reduce our unsustainably high levels of solid-waste creation, for example, it is more effective to design a circular economy, which reuses and recycles as much waste as possible, than to consume less by buying fewer products that will end up in the waste stream. For animal agriculture, a consumption-based approach makes no sense: people have to eat in order to survive, so reducing consumption of food, measured in calories, won't work. But encouraging and incentivizing a plant-based diet will gradually reduce the impacts of eating meat.

Nuclear power

There is a lot of debate in the environmental community about whether nuclear energy should be part of the solution for climate change. It's not really a renewable resource—it uses uranium as fuel, and there's a limited supply. But there's enough uranium left to power reactors for hundreds of years. Generating electricity in nuclear plants releases almost no greenhouse gases. But concerns about safety and the disposal of radioactive waste still linger. Newer technology, which the industry touts as much safer, has been developed over the last few decades, and may alleviate these concerns. But it takes many years for a new nuclear-power plant to

---

[45] O' Neill and et al., 92.

be approved and built, at least in the US. So the question becomes: given the long delay in bringing a new plant online, can nuclear power play a significant role in getting us quickly to the point where we're no longer burning fossil fuels? It's a tough policy choice, and there are strong arguments on both sides of the issue.

Land use planning

A majority of Earth's population lives in cities now, and an additional 2.5 billion people will live in cities by 2050, with up to 90 percent of the increase concentrated in Asia and Africa.[46] Generally speaking, the way we use land is planned in cities, but not in rural areas. Urban land use has a huge effect on GHG emissions, mostly by way of buildings and transport. Models cited by the IPCC show that urban GHG emissions, currently 29 Gt/year, could rise to 40 Gt in 2050 if mitigation efforts are low, but could be reduced to 3 Gt/year with a high level of mitigation.[47]

I work as an environmental and land-use attorney in Los Angeles. Most of this work is litigation against development projects, either housing or warehouses. These cases give me a ringside view of the planning process conducted by the City and County of Los Angeles. I fight warehouses mostly to compel them to fully mitigate their GHG emissions; there aren't many planning issues for warehouses. I litigate against housing developments partly to compel them to reduce GHG emissions. Litigation I was involved in against two of the largest new cities of around 20,000 homes forced those projects to become net-zero projects by reducing on-site GHG emissions as much as possible, and then mitigating the remaining emissions off-site, through local projects and offsets.

I also fight housing projects that are urban sprawl. I'm currently suing Los Angeles County over a 37-house project on 94 acres north of the City of Los Angeles, to be built in what is currently an open-space canyon. I also sued over Five Point Valencia (formerly Newhall Ranch), a new city of 20,885

---

[46] IPCC WGII, "Climate Change 2022: Impacts, Adaptation and Vulnerability," 909.
[47] IPCC WGIII, "Climate Change 2022: Mitigation of Climate Change" (IPCC, 2022), 30, https://www.ipcc.ch/report/ar6/wg3/.

dwelling units housing 58,000 residents plus commercial and business establishments, schools, golf courses, parks, etc. One part of the project, called "Landmark Village," is to be built largely in the floodplain of the Santa Clara River, the last free-flowing river in Southern California. The floodplain is about a mile wide. The river is seasonal, with big flows only after major rainfall. The channel where the water flows is much narrower than the floodplain; it moves around within the floodplain as erosion and other forces change its path.[48] Rivers are special and important ecosystems, and this river provides habitat very different from the desert chapparal that surrounds it. From an environmental perspective, the river floodplain is the worst place to put a new city. To prevent flooding in the new city, the river must be "channelized," by constraining the flow within a narrow concrete channel passing through the development. Channelization totally changes the ecology of the river, effectively converting it into a narrow concrete storm drain.

The Los Angeles River, which flows through the city of that name, and its tributaries, which cover most of the city, were converted into concrete storm drains by the Army Corps of Engineers after extensive flooding occurred in 1914, 1934, and 1938.[49] Local government, having finally realized that the river is a valuable natural and social resource, is now spending over a billion dollars to restore the river to a more natural condition. It would have been much better to have altered the landscape in a more natural way, instead of the concrete channelization, last century. But that would have required much more land to be kept out of the housing market, and the demolition of some existing housing.

Channelizing these two rivers was a gross planning error based on short-term economics and politics. Five Points is a huge landowner with deep political connections. They wanted to develop the land they own in the Santa Clara River. For them, there was no question of moving the project to a more suitable location, because they didn't own any such real estate.

---

[48] Luna B. Leopold, *Waters, Rivers and Creeks* (Sausalito, CA: University Science Books, 1997), 59–116.
[49] Blake Gumprecht, *The Los Angeles River* (Baltimore: The Johns Hopkins University Press, 1999), 173–233.

So the County approved their proposal due to their political clout. Similarly, when the Army Corps built flood control systems to protect Los Angeles from flooding, they were constrained by existing housing and other buildings located adjacent to the river channel. To preserve these improvements, they needed to build a narrow channel instead of constructing a wider and more natural landscape for the river. This is the typical pattern: exceptions to urban planning rules are granted all the time, to solve short-term problems, or to accommodate the short-term needs of well-connected real-estate developers.

The planning system is supposed to work a different way: The local government (city or county) develops a "general plan," a land-use plan for how land in the local jurisdiction will be used. It contains sub-plans for traffic, housing, conservation, open space, noise, safety, and environmental justice. But the most important part is the land-use element, which designates uses (residential, commercial, industrial, park, open space) for each area of land within the jurisdiction. Zoning is adopted in conformance with the land-use designations; it specifies in more detail which exact uses are permitted in each type of zone. A person wishing to build something on his or her land must build in conformance with the general plan and zoning. For example, he or she can't build an apartment house on land designated for a single-family house.

What happens in practice in Los Angeles is that smaller landowners conform to the zoning and land-use designation, but large landowners have the political capital to get the City Council to change the plans. For both of the large projects I fought, the landowner got a "specific plan" approved by the County Board of Supervisors to change their properties from agriculture and open space to new designations and zoning conforming with the projects they want to build, effectively bypassing the planning system.

Traditional, single-use zoning has become passé in urban planning circles. It physically segregates the various types of uses. Residential zones contain only housing, and commercial zones contain just stores, with no housing. Walkable, dense, mixed-use neighborhoods, well connected to transit, provide, for many, a better living experience than traditional single-family houses on large lots in residential-only neighborhoods. Dense mixed-use neighborhoods can also greatly reduce GHG emissions. One way they do this is by reducing automobile vehicle miles travelled (VMT). In a walkable neighborhood some trips can be made on foot or bike. Transit is much more likely to be utilized by those who can walk or bike to it, instead of having to drive to the transit terminal. And higher density means that many destinations will be physically closer so that trips that need to be made by auto are shorter. Some cities, such as Los Angeles, have started to change their zoning codes to allow mixed-use transit-oriented developments.

Planning at the regional level is vital, especially for transportation. Cities contain centers that need to be linked to one another and to other cities by transit and roads. The traditional pattern of single-family houses in suburbs designed for automobiles, with many residents commuting into the city by car, should be eschewed in favor of regional centers and neighborhoods with jobs and homes close to transit. We should stop expanding our roads and highways and put the money into developing public transit instead. Regional planning efforts, in cooperation with local (city and county) government, should control the urban growth boundary, prohibiting sprawl and encouraging infill development.

Peter Calthorpe, in his excellent book, *Urbanism in the Age of Climate Change*, suggests that "urbanism," involving the design principles I've just been describing, can make a huge contribution to curbing global heating. Urbanism may be an important long-term strategy, especially for the new cities that will be built out in Africa and Asia, but it won't happen quickly enough in mature cities to significantly reduce GHG emissions by 2050. The planning process just takes too long, probably decades from the time the process is started until it materially affects actual buildings and infrastructure that are being constructed.

Building codes should be changed as soon as possible to require new construction to have net-zero GHG emissions. Since we will need to phase out natural-gas and fuel-oil use when we stop burning fossil fuels by 2050, we should not allow these fuels to be used in new construction. Passive solar design, solar water heating, solar panels to generate, and batteries to store, electricity should be required. Effective insulation and weatherstripping should also be required. These same rules need to be applied to existing buildings; we'll miss many of the benefits of net-zero construction if new buildings are the only ones that are net-zero. But, even in California, there is no public program to retrofit existing buildings. We need to change that and provide financial incentives for retrofits to building owners.

SER framework

The AR6 WGIII Report refers to the Sufficiency-Efficiency-Renewables (SER) framework.[50] The idea behind "sufficiency" is that we don't really need everything we're consuming. For example, many of the new single-family houses being built in Los Angeles now are larger than 400 m², when most families could live happily in half that much space. "Sufficiency is defined as avoiding the demand for materials, energy, land, water and other natural resources while delivering a decent living standard for all within the planetary boundaries."[51] This is similar to the "avoid-shift-improve" framework proposed by economists.[52] Examples of avoidance/sufficiency are:

- teleworking, to avoid the resource costs of commuting,
- making cities more compact and walkable, to avoid the resource costs of mechanical transportation,
- passive heating and cooling, to avoid the resource costs of heating and air conditioning,
- longer-lasting appliances, to avoid the resource costs of producing new appliances, and
- reducing calorie intake to what's required, to avoid the resource costs of producing food and disposing of food waste.

Avoidance usually has a negative cost.

---

[50] IPCC WGIII, "Climate Change 2022: Mitigation of Climate Change," 957
[51] IPCC WGIII, 957.
[52] Felix Creutzig and et al., "Towards Demand-Side Solutions for Mitigating Climate Change," *Nature Climate Change* 8 (2018): 260–63.

Efficiency also has a negative cost, at least in the long run. For example, replacing incandescent light bulbs with LED bulbs requires a small short-term investment, but that investment is repaid many times over the long life of the LED bulb by electricity savings, resulting in a long-term negative cost.

UN sustainable development goals

In 2015, the United Nations held a Sustainable Development Summit,[53] which finalized the 2030 Agenda for Sustainable Development,[54] which was adopted by all UN Member States. It specifies 17 Sustainable Development Goals (SDGs) to be achieved by 2030, relating to the three dimensions of sustainable development it sets forth: economic, social, and environmental. It's a fairly comprehensive effort, including mostly social and economic goals such as ending poverty and hunger and ensuring healthy lives and quality education for all. But it includes sustainability goals for the environment. Goal 13 is to "take urgent action to combat climate change and its impacts." Goals 14 and 15 are to protect and preserve oceans and land in a sustainable manner.

It's a laudable effort, and it would be wonderful to achieve these goals, but there's no real commitment among the countries who signed onto the goals to achieve them; they're aspirational. We'll see this a lot when we look at international law in the next chapter; there are a lot of "soft laws" containing aspirational goals, with no mechanism to insure they are implemented.

Conclusion

What should we take away from this discussion of sustainability? Here are the main points:

---

[53] United Nations, "U.N. Sustainable Development Goals," 2015, https://sdgs.un.org/goals.
[54] United Nations, "2030 Agenda for Sustainable Development," September 27, 2015, https://sdgs.un.org/2030agenda.

- Exponential increases are not sustainable and will not continue long-term. This means, among other things, that the high level of economic growth we've been experiencing since World War II will not continue for hundreds of years more.
- Population × consumption is not a good measure of humans' environmental impacts on the Earth. Earth's population will probably level off at a sustainable level by the end of the century, and the impacts of consumption depend on what is consumed.
- As developed countries continue to grow their economies and developing countries develop stronger economies, we must ensure that we are not using Earth's resources at a rate that is unsustainable.

Chapter 4
Economics

Introduction

Talking about climate with me, a fairly wealthy friend suggested, a few years ago, "why not just do a cost-benefit analysis?" This implies that the issue should be viewed primarily through an economic lens. That makes some sense for my friend and me—the impacts for us will be primarily economic. The temperature where I live in Los Angeles hovered around 46°C for a couple days two summers ago. We're experiencing a drought and water shortage, so we can't water our plants as much as they need. And we were threatened by wildfires from across the street three or four years ago. But my friend and I have enough money to adapt, with air conditioning, insurance, and some changes in lifestyle. A lot of people in this world, however, don't have sufficient economic resources to adapt, and won't experience the harms that global heating causes them as economic harms but rather as diminutions in their already low life circumstances.

As I argued in Chapter 1, the top-level framework for a proper discussion of climate impacts, must be ethics, not economics. An economic analysis is convenient because it reduces everything to money, which we can deal with quantitatively, using a large body of economic theory. And climate change will manifest itself to wealthier people largely as an economic issue. But there are many impacts that will be primarily non-economic, such as the loss of our biodiversity and harm to natural resources such as forests.

The weather where rich people live may get worse (or better!), they may be affected by water shortages, droughts, increased storms, heat waves, sea-level rise, etc., but they have resources to adapt. They can afford to move to a place with a better climate, if need be. Poor people may not have these options, and may suffer more of these direct climate effects, as well as economic impacts.

But the main reason we can't use economics as the top-level frame of our discussion of climate ethics is that there are many things important to people that are difficult to value properly in economic terms. Human life and health are notorious in this respect.

For legal purposes, future lifetime earnings are often used to value lives. They were used, for example, to compute damages for the deaths in the Twin Towers attacks of 9/11/2001. Some of the difficulties are illustrated in a 1967 paper by D.P. Rice and B.S. Cooper on "the economic value of human life."[1] That paper contains tables showing that projected future lifetime earnings of white, college-educated males, between 30 and 34 years of age are $223,471, while projected earnings of women of color with less than 8 years of school max out at $54,417. Valuing lives this way is racist, sexist, and callous.

But the bigger problem is the assumption that a monetary value can be put on human life. The two are incommensurable. Economists often use "willingness to pay" to value goods or services for which there is no market. It can be estimated by taking a survey. Most people I know would be willing to pay a lot for an extra year of life, especially if it were immediate (you die tomorrow if you don't pay today) and not just tacked onto the end of a presumably long life. And, in this thought experiment, do we allow bids that exceed the bidder's net worth? Could I be "willing" to pay a billion dollars to prolong my life, even though I have no practical way to obtain this much money? These problems with willingness to pay show how incommensurable are economic value and the value of human life.

And there are a lot of things I value other than my life and health, that are not readily valued by money. What are the damages to me from ecological harms to Yosemite National Park, which I usually visit each year? How much does it harm me when the Earth loses a plant species? How do we arrive at a monetary damage figure for the increased temperatures, which make the place where I live less pleasant?

---

[1] D.P. Rice and B.S. Cooper, "The Economic Value of Human Life," *Am J Public Health Nations Health 57*, no. 11 (1967), https://pubmed.ncbi.nlm.nih.gov/6069745/.

If we allow that non-human animals have their own moral standing, and are not just instrumental to humans, how to we value the damages that global heating will do to them? They don't participate at all in the economic system, and don't have or care about money. And, as we shall discuss below, economics is unsatisfactory in dealing with future generations that will be harmed by climate change.

Even though we cannot use economics as the top-level frame for our discussion of how to deal with global heating, they are of great value in helping us with certain parts of the problem. We can use economic tools to decide how to make trade-offs between options that can be expressed in economic terms, such as how much mitigation is optimal.

Sources

There are two important books on climate economics by well-regarded economists. The first is *The Climate Casino*,[2] which won a Nobel Prize for its author, Yale economist William Nordhaus. The second is *The Economics of Climate Change: The Stern Review*[3] by Nicholas Stern, the Chair of the Grantham Research Institute on Climate Change and the Environment at the London School of Economics, an advisor to the UK government, and the former chief economist of the World Bank. The IPCC AR6 WGIII Report[4] also has a lot to say about climate economics.

An economic analysis framework

Total climate cost = climate damages + mitigation cost + adaptation cost

I'm going to use the above framework for the economic analysis in this chapter. Climate change is causing increased storms, droughts, and

---

[2] William Nordhaus, *The Climate Casino: Risk, Uncertainty, and Economics for a Warming World* (New Haven: Yale University Press, 2013).

[3] Nicholas Stern, *The Economics of Climate Change: The Stern Review* (Cambridge, UK: Cambridge University Press, 2006).

[4] IPCC WGIII, "Climate Change 2022: Mitigation of Climate Change" (IPCC, 2022), https://www.ipcc.ch/report/ar6/wg3/.

wildfires, for example, and the resulting harms can be measured, though incompletely, in monetary terms. Those are examples of climate damages. In evaluating those damages, we're concerned with the difference between the pre-climate world where the harms didn't occur and our current and future worlds, where climate harms are the norm. Damages affect different people in different ways and this can heavily affect their valuation. For example, a river overflowing due to a heavy rainstorm in Bangladesh may cause far less monetary damage than the same level of overflow in a river in Europe, though the human misery caused by the overflow may be much higher in Bangladesh.

As discussed in a Chapter 2, mitigation is action we take to reduce the causes of global hearing, such as reducing our burning of fossil fuels. The usual metric for mitigation costs is US Dollars (USD) per metric ton of avoided CO2e emissions (tCO2e). Costs tend to range from $5 to $1,000 per tCO2e. Some mitigations even have negative costs. Conserving energy by replacing incandescent lightbulbs with LEDs, for example, saves money over the lifetime of the bulbs (i.e. has a negative cost), but qualifies as mitigation because it lowers energy use and reduces GHG emissions from electricity generation. It's important to consider the cost of various types of mitigations because we usually want to implement policies leading to the cheapest mitigations. If we keep doing this, then mitigation costs per tCO2e will gradually rise as we do more mitigation.

Mitigation is universal and not particularized like damages. As far as its effect on the climate system, a tCO2e avoided by energy conservation has the same effect as a tCO2e absorbed by a growing forest. Mitigation mostly consists of stopping the process of emitting GHGs, mostly by stopping the burning of fossil fuels. There are other types of mitigation, such as reforestation, which grows new forests to absorb CO_2 from the atmosphere.

Adaptation is the process of adjustment to actual or expected climate and its effects in order to moderate harm. Some examples of adaptation are building a sea wall to armor a coastal building against sea-level rise, installing air conditioning to protect against heat waves, and procuring alternative sources of water for areas afflicted by drought.

The relative proportions of these three costs depends on the amount of warming we allow before achieving carbon neutrality. Limiting temperature increases to 2°C would greatly reduce climate damages, but would require more money to be spent immediately on mitigation. Allowing an increase of 4°C would reduce mitigation costs, but would open us up to increased damages. There is a theoretical economic optimum which minimizes the sum of damages, mitigation and adaptation costs and this points to the best course of action, from a strictly economic point of view. The graph below illustrates this concept, using made-up data. Limiting temperature increases to a small amount incurs high mitigation costs, but allowing temperatures to increase to high levels incurs high climate damages. The economic optimum, minimizing total costs, which are the sum of damages (and adaptation costs) and mitigation costs, occurs in the graph below around 2°. Limiting warming to a level lower than 2° incurs high mitigation costs; allowing warming greater than 3° results in high damages and adaptation costs.

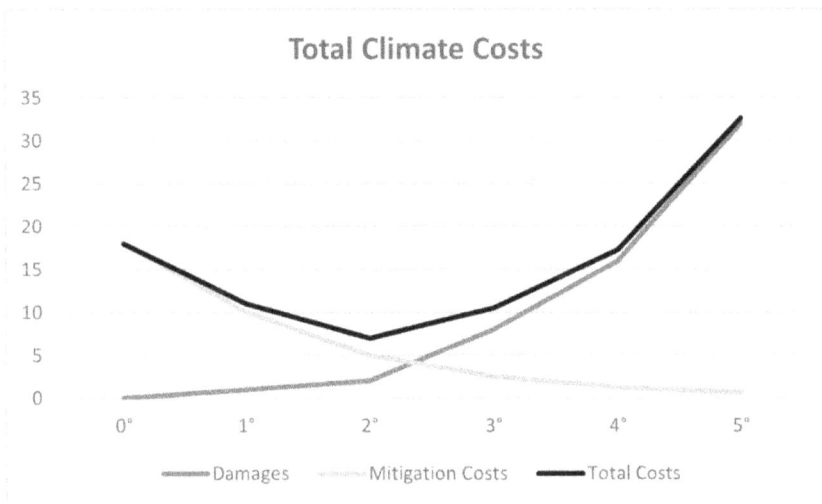

Figure 4-1: *Total Climate Costs*

GDP

Most economists estimate climate change-related costs as a percentage of global gross domestic product (GDP), estimating total climate costs in a range from about 1% to 10% of GDP. What does this mean? GDP is total income for all persons in the jurisdiction: it is the sum of employee compensation, rents, interest, proprietors' income from business they own, and corporate profits.[5] All of these are personal income, including corporate profits, since corporations are owned by individuals. The GDP of Germany, for example, is the total income received by all persons in Germany. It is also the total value of all goods and services produced in Germany.

When an economist estimates that something will cost 3% of GDP in some particular year (like 2050), what does that mean? As I write this, total worldwide GDP is $84.7 trillion and, since the global population is about 8 billion people, the average per-capita GDP is $10,558. So, if global GDP were reduced by 3%, the average person would lose $317 in the year in question. This is a global average and, because of extreme income inequality, there is a huge spread in personal income, from those making less than USD $2.15/day (the World Banks's extreme-poverty level) up to the ultra-rich, who have annual incomes in the billions of dollars. Most people should assess the 3% cost as costing 3% of their income, so if they're relatively well off and make USD $100,000 per year, it would cost them $3,000.

Economists sometimes express climate costs in terms of a diminution of economic growth, which is usually calculated based on GDP. Global GDP grew at an average of around 6% per year over the last 60 years,[6] though it slowed to 2.8% between 1990 and 2019.[7] So if climate costs are 3% of GDP, the costs amount to a year's economic growth, and can be viewed as costing a year of growth. I prefer to decouple the discussion of climate costs from economic growth because, as discussed in the previous chapter on sustainability, we need to reconsider the value and necessity of growth, particularly the components of growth that use increasing Earth resources.

---

[5] Campbel R. McConnell, Stanley L. Brue, and Sean M. Flynn, *Macroeconomics*, 21st ed. (New York: McGraw-Hill Education, 2018), 144.
[6] World Bank, "GDP Growth," 2022,
https://data.worldbank.org/indicator/NY.GDP.MKTP.KD.ZG.
[7] IPCC WGIII, "Climate Change 2022: Mitigation of Climate Change," 3-24.

A related concept is the Social Cost of Carbon (SCC), which is the amount of net damages that will be caused by the emission of one more ton of CO_2e.[8] SCC is useful for considering policy alternatives from an economic point of view. The US government recently established SCC values for carbon dioxide at \$51/ton, for methane at \$1,500/ton, and for nitrous oxide at \$18,000 per ton.[9] The SCC depends heavily on what happens in the future. Scenarios with little mitigation result in high SCC estimates because future climate damages will be very high; scenarios with early and robust mitigation result in lower SCC estimates. Discounting (discussed below) also plays a big role. A 2018 study estimated a global median SCC of USD \$417/ton under middle-mitigation scenarios.[10] The place where GHGs are emitted doesn't affect the SCC, but there are big regional differences in impact. For some cold countries such as Canada and Russia, the SCC may be negative, meaning that an incremental increase in GHG concentrations, and a corresponding incremental increase in temperature, may be beneficial.[11]

Externalities

Environmental harms are usually ignored by markets—they are not included in the prices of goods and services. Such harms (or benefits, when they are positive) are called externalities. For example, when I burn a gallon of gasoline by driving my car, I am releasing about 20 pounds of CO_2 into the atmosphere. Burning 110 gallons releases a ton of CO_2. If the SCC is \$51/ton, it amounts to about 45¢ per gallon. But this cost is not included in the price and therefore does not affect the market pricing of gasoline. We could correct this situation by collecting a carbon tax of 45¢/gallon of

---

[8] IPCC WGII, "Climate Change 2022: Impacts, Adaptation and Vulnerability" (IPCC, 2022), 16–115.

[9] Rachel Cleetus, "The Social Cost of Carbon Gets an Interim Update from the Biden Administration," *The Equation* (blog), March 2, 2021, https://blog.ucsusa.org/rachel-cleetus/the-social-cost-of-carbon-gets-an-interim-update-from-the-biden-administration/.

[10] Katharine Ricke et al., "Country-Level Social Cost of Carbon," *Nature Climate Change* 8, no. October 2018 (2018): 897.

[11] Ricke et al., 897.

gasoline at the pump. Such a tax would induce consumers to use less gasoline, by driving less or buying more fuel-efficient cars. The taxes could be used to fight climate change, such as by funding mitigation research.

Many other environmental impacts are externalities, such as the costs of air pollution from industry, power plants and cars. Another example is the cost added to our healthcare system, born by all of us, from smoking tobacco or refusing to get vaccinated against COVID.

Paths to net zero

As discussed in Chapter 2, we need to reach net-zero GHG emissions as soon as possible, to keep temperatures from rising further. The scientific consensus is that we must achieve net zero by 2050 in order to keep the maximum global temperature increase around 2°C. Net zero means that we allow fossil fuels to be burned only to the extent that we remove an equivalent amount of GHG emissions from the atmosphere. Right now, there is no technology that can remove CO_2 at the required scale cost-effectively, so we need to plan for phasing out the production and consumption of coal, oil, and natural gas by 2050.

There are two types of policies that can be used to accomplish this goal. The first uses market mechanisms, such as carbon taxes and cap-and-trade programs. For example, we could tax GHG emissions, starting at the current SCC, and gradually increasing the tax per emitted ton to achieve net zero by 2050. The second type of policy is what economists call, disparagingly, "command-and-control" regulation. We could legislate now that it will be illegal to burn fossil fuels without a permit (which would require extraordinary circumstances to obtain) in 2050. A country enacting such legislation would show that it is serious about achieving net zero, and would send a strong signal that everyone must conform, over the next 25 years, to the net-zero reality.

Even though economists deprecate command-and-control regulation as a policy instrument, a law prohibiting the burning of fossil fuels starting in 2050 would still achieve most of its results via market mechanisms. The oil majors would be certain that their current business model would not continue past that date, so they would stop building infrastructure such as pipelines, and stop exploring for new sources of oil and gas. Consumers purchasing cars would know that gasoline- or diesel-powered cars would become worthless on that date. Home buyers would prefer all-electric homes, knowing that any gas- or oil-powered heaters would need to be replaced by 2050.

We can illuminate some of the tradeoffs by considering what lawyers would call a "hypothetical," and scientists a "thought experiment." What if we prohibited the burning of fossil fuels right now? About two-thirds of our electricity-generating capacity would need to be shut down and replaced with renewable generation. 97% of the vehicles on our roads would need to be scrapped and replaced with zero-emissions vehicles. Central heating systems in most buildings, which are now powered by heating oil or gas, would need to be shut down and replaced with electric heat pumps. Almost all manufacturing of cement for concrete would be halted, until research is done and new processes for producing Portland cement are developed. Almost all aviation would be grounded until a new circular process was developed for producing net-zero aviation fuel. Market forces would optimize our response to such a prohibition, making appropriate trade-offs among the options, in most cases.

Stopping GHG emissions now would keep global warming to a minimum. GHG concentrations would stop increasing, as would the global temperature, except for, possibly, some committed warming as the atmosphere equilibrates with the ocean. This hypothetical illustrates the leftmost achievable point in the Global Climate Costs graph above. Climate damages and adaptation costs are minimized, but the mitigation costs would be astronomical. But if we ignore the situation and allow temperatures to increase without limit, the climate damages will be even more astronomical.

All of the actions we would need to take if we stopped emitting GHGs now will need to be taken, on a global basis, before 2050, but we're not acting like we're aware of this huge and expensive to-do list. Economics can help us prioritize the less-expensive mitigations (including those with negative costs, such as energy conservation) so we do them first.

Costs vs. investments

Under standard accounting rules for businesses, expenditures can be either costs or investments. Costs are immediate expenses, such a money spent for salaries, rent, or supplies. But when a business spends money to acquire an asset that will be used long-term, such as an electricity-generation plant or a pipeline, that is a capital expenditure, an investment. It does not count as an expense in the short term. Instead, the acquired asset goes on the balance sheet as a long-term (non-current) asset, which is depreciated each year over the lifetime of the asset, until, at the end of the lifetime, the value becomes zero. The yearly depreciation is the expense that is reflected on the income statement, subtracted from business revenues to calculate profit. The norm is that, when a business acquires a capital asset, it exchanges one asset (cash) for another (the asset), and this even exchange of assets does not reduce corporate profits. The depreciation every year over the lifetime of the asset, is an expense reflecting the cost of using the asset during the year and it is charged against corporate profits for the purpose of determining net income.

If an asset is taken out of service before the end of its useful life, it becomes a stranded asset, and the portion of its value that has not yet been depreciated becomes an immediate expense. For example, if a power plant with a 20-year lifetime initially cost $1 billion, and if its owner used straight-line depreciation, it would expense $50 million per year for the use of the plant. But if the plant were taken out of service halfway through its lifetime, $500 million of its value would still remain on the books and would need to be written off as an expense at the time the asset was taken out of service. Such costs for stranded assets will form a significant fraction of global-warming mitigation costs.

Conversely, if a business is building a new (presumably renewable) electricity-generation plant, the cost on the company's books will be spread out over the lifetime of the plant. But the company will need to fund the investment in the plant, using retained earnings, borrowing the money, or finding an outside equity investor. Such investments are risky, and can be difficult to obtain. Investors may prefer less risky projects with higher rates of return. The lack of investment capital is an important impediment to adequate climate mitigation world-wide. The latest IPCC report estimates that climate financing needs to increase by a factor between 4 and 8 in developing countries, and by a factor between 2 and 5 in developed countries.[12] This is a huge gap. Government policies could help reduce this gap; for example, governments could guarantee loans, or provide tax incentives for climate finance.

Economic growth

The previous chapter on sustainability touched on economic growth, but a fuller discussion here will be useful for understanding the economics of climate change. Growth is usually measured in terms of GDP: it is the increase (or decrease) in GDP from one year to the next, usually expressed as a percentage. Optimists expect GDP to keep growing, so that each generation will be wealthier than the generation, and they expect global GDP to balloon as developing nations raise their standards of living. I argue in this section that we can't rely on GDP continuing to grow at the high rate of the recent past.

Economic growth is generally a good thing. Before the industrial revolution, which started around 1776, there was little economic growth—living standards had been more or less the same for two thousand years. This reminds me that the same was true for the technology of war. When I see the Battle of Agincourt (1415) depicted in a movie, the fighting—with swords and spears, the elite on horses, others on foot—is very similar to the fighting in Homer's Iliad, which describes events that supposedly took place three thousand years ago. But, starting around 1776, when James Watt perfected the steam engine, we started to use machines to increase labor productivity. US per-capita GDP increased from $2,100 in 1800 to $69,288 in 2021. In materials terms, we live a lot better than the Americans in 1800, due to our increased consumption. We also live a lot better than the residents of India, where per-capita GDP is still just $1,980.

---

[12] IPCC WGIII, "Climate Change 2022: Mitigation of Climate Change," 1577.

But increased damage to the environment comes with increased GDP.[13] GDP can be broken down as population × per-capita consumption. Population growth increases environmental harm, as discussed in the sustainability chapter. That chapter made the case that we need to control population and, ideally, limit the population to an agreed-upon level. If the population stopped growing, we could stop building out our cities, especially in developed countries. In Los Angeles, where I live, we're still building more housing out into open space surrounding the city. State laws mandate that cities and counties plan for building a lot more housing, even though the state's population is slightly declining.

The situation is different in developing countries, some of which are still experiencing strong population growth. They hope to become affluent like the developed countries, and will increase their per-capita GDP to do so. We need to help them do this in an environmentally responsible manner.

GDP growth harms the environment when it results in an unsustainable level of resource use. Energy, materials, and land are the most significant resources. Energy intensity—the amount of energy used per dollar of GDP—is declining slowly, a percent or two per year.[14] Energy is the main focus of climate policy. There are two components to reducing GHG emissions from energy: (1) conservation so we use less energy and decrease energy intensity of the economy; and (2) switching from fossil fuels to renewable-energy sources, so the energy we use does not contribute to global heating.

We are using other resources at an unsustainable level. Mining, deforestation to support agriculture and grazing, oil production to obtain raw materials for plastics, agriculture, meat production, and grazing are all using environmental resources at un unsustainable rate, at least in some

---

[13] James D. Ward et al., "Is Decoupling GDP Growth from Environmental Impact Possible?," *Plos ONE* 11, no. 10 (2016), https://journals.plos.org/plosone/article/file?id=10.1371/journal.pone.0164733&typ e=printable.

[14] International Energy Agency, "Energy Efficiency 2020," 85, https://www.iea.org/reports/energy-efficiency-2020.

places. And, in many places we're drawing too much from groundwater and surface water. Many of these environmental impacts will be intensified by global heating. For example, water shortages will be greatly exacerbated. So we need to reduce our unsustainable resource use as part of our response to climate change.

When accounting for GDP from the production side, wages are the most important component of GDP. (Other components are rents, interest and profits.) Wages can be broken down into hours worked × labor productivity. This means that GDP can be increased either by increasing the number of hours worked (by increasing the population), or by increasing labor productivity.

The main impact of the industrial revolution has been to increase labor productivity, the value of an hour's labor. Better tools, technologies and production methods, using computers and robots, cause productivity to increase. Such increases in labor productivity, instead of population increases, should become the driving force for economic growth in the future. Labor productivity is currently growing at 3.2% among OECD (Organisation for Economic Co-operation and Development—relatively developed) countries.[15]

However, labor productivity is increased by innovation resulting from research and development (R&D). The productivity of R&D, in terms of the value of the innovations per researcher (research productivity) has been declining at a rate of around 5% per year for a long time.[16] For example, in order for Moore's law, which posits a doubling of transistor density on computer chips every two years, to continue, eighteen times as much research effort is required now for each doubling than was required in 1971, when the first computer chips were developed. This is true healthcare and agriculture as well, and seems to be generally true across the economy. In

---

[15] OECD, "Labour Productivity and Utilisation" (OECD, 2021), https://data.oecd.org/lprdty/labour-productivity-and-utilisation.htm#indicator-chart.

[16] Nicholas Bloom et al., "Are Ideas Getting Harder to Find?," *American Economic Review* 110, no. 4 (2020): 1104–44, https://doi.org/10.1257/aer.20180338.

order to keep innovating at the same rate, we therefore need to increase our research expenditures by 5% per year, and we can't continue doing that forever because it is an exponential increase. So we are unlikely to be able to continue our current high rate of economic growth for more than a decade or more in the future; falling R&D productivity will lead to smaller increases in labor productivity, and lower economic growth.

Different cultures and countries have different approaches to work. Western Europeans, on average, work fewer hours and therefore receive less income than workers in the US. They've decided to trade some of their potential income for more leisure time. For example, Germany's 2021 per-capita GDP was 36% less than that of the US, and Germans worked 33% fewer hours, on average. Working less reduces GDP and economic growth, but it may increase well-being.

One of the arguments in favor of deferring action and expenses on climate is that, because of economic growth, future generations will be richer than we are, and more able to pay. But we should be wary of relying too much on a continued high rate of growth, for several reasons. First, we cannot expect much further population growth. Second, we need to stop using Earth resources more than is sustainable, another GDP reduction. Third, we may need a massive transfer of resources from developed to developing countries, to allow them to grow their economies and become prosperous without greatly exacerbating global heating. Fourth, our exponential increase in R&D expenditures, which is driving GDP growth, can't continue forever.

Discounting

As discussed in previous chapters, inter-generational equity is one of the most important ethical issues in dealing with climate change. Discounting is the way economists deal with this problem. As with economics in general as a framework for solving climate problems, discounting is insufficient for our purposes, but it is important to understand it, since it is one of the biggest factors in modelling climate impacts and responses. Discounting is an attempt to deal with an important trade-off between the present and future: our expenses to mitigate climate damages start now and will continue, but the benefits from mitigation will accrue gradually, and extend into the next few hundred years.

Discounting is the way economists answer the question: "How much is a future expense or income worth today?" For example, we could ask "how much should I be willing to pay today in exchange for a guaranteed payment of $100 ten years from today?" We'll make simplifying assumptions, such as ignoring the risk that I won't get paid in ten years because the person who's supposed to pay me is bankrupt. If I know I can get a 3% annual return on my money over ten years, I can calculate the amount of money I'd have to put into an interest-bearing account today in order to have $100 in ten years. That amount is $74.40. This amount is the net present value (NPV) of $100 ten years from now. Three percent is the discount rate. And the process of calculating the NPV is called discounting.

The same calculation can be used to discount expenses. The NPV of an expenditure of $100 in 10 years, at a 3% discount rate, is $74.40. This says, based on economics, we should be willing to pay $74.40 today in order to avoid an expense of $100 in 10 years. Discounting is frequently applied to regular streams of income, like mortgage payments. There is a simple formula to calculate NPV of a perpetual stream of regular expenses: Y/r, where Y is the yearly amount and r is the interest rate, expressed as a fraction.[17] The NPV of an income or expense stream of $100/year forever at a 10% discount rate is $100/0.1 = $1,000. The same stream, at a 1% discount rate is worth $100/0.01 = $10,000 now. With no discounting (i.e. r=0), the NPV is infinite; this makes sense because it represents the sum of an infinite number of future $100 receipts or payments that are all deemed to be worth $100 now. When discussing discounting, economists usually adjust for inflation.

We need to dig a little deeper into discounting, to apply it to climate economics. Here are some of the factors that go into determining an appropriate discount rate:

---

[17] McConnell, Brue, and Flynn, *Macroeconomics* (New York: McGraw-Hill Education, 2018), 90.

- Current rate of return on investment
- Economic growth rate
- Inflation
- Risk
- Human desire to have benefits now rather than later
- Intergenerational equity

The rate of return on investment is usually the starting point for determining an appropriate discount rate because it determines how much a given investment made today will be worth at some point in the future. The GDP growth rate is also relevant, because it determines how much richer than us future generations will be. We'll assume for this analysis that all figures have been adjusted for inflation, so we won't factor it in separately. Risk generally increases the discount rate, because investors want a higher rate of return when they might not receive their future payout. And there is a natural human desire to get benefits now rather than later, which economists call the "pure rate of time preference."

Finally, the big one, intergenerational equity. This is not usually taken into account when deciding on a (non-climate-related) discount rate because most situations, such as the stream of payments made for a home mortgage, don't extend multiple generations into the future. A zero discount rate treats future persons the same as current persons. This is philosophically satisfying, but impractical; it makes the NPV of future climate costs infinite, or at least extremely high. On the other hand, discount rates of five percent or higher, which have been proposed, are a way of saying "Don't worry, let the future take care of itself. We'll have so much more money and so much better technology in the future that we'll be able to solve the problem with little economic harm." The reverse precautionary principle, in other words.

Nordhaus' argument for a 4% discount rate is based on a 4% real rate of return on invested capital. He points out that there are many productive investments we could make, such as healthcare research or retraining our workforce, and our basis for choosing among the investment policies should be the return on invested capital. This is a valid point, but it also illustrates why ethics, instead of economics, should be the main framework for evaluating the world's response to global heating. As discussed above, climate change will result in many impacts that are not addressed by economic markets, such as the loss of biodiversity, and the harm to our national parks and other lands we preserve in close to their natural state. And many climate impacts on the poor are undervalued by traditional economic analysis, because the poor participate in the economy at such a low level and are often not compensated for climate damages.

The Stern Review uses a discount rate of 1.4%, based on a 1.3% rate of per-capita consumption growth,[18] plus 0.1% for pure time preference. This latter low rate is essentially an inter-generational rate of 0%, tempered slightly by the small possibility that the human race will be extinguished by an asteroid or some other existential impact.

I'll advocate for a 2% discount rate, following the results of a 2018 survey of 200 economists working in this area[19] and the IPCC WGIII's finding that 2% is supported by the majority of experts.[20] In the survey, recommended discount rates ranged from 0 to 10 percent; 92% of the experts said they would be comfortable with a discount rate between 1 and 3 percent, and over three-quarters found a discount rate of 2% acceptable. This contrasts with Nordhaus' 4% discount rate,[21] and Stern's 1.4%.

"The modelled cost-optimal balance of mitigation action over time strongly depends on the discount rate used to compute or evaluate mitigation pathways; lower discount rates favour earlier mitigation…"[22] Discounting shifts the optimum discussed above because it reduces the present value of climate damages, which will occur in the future. A lower discount rate increases the net present value of future damages and shifts the optimum towards quicker action to mitigate climate damage.

---

[18] Stern, *The Economics of Climate Change: The Stern Review* (Cambridge, UK, Cambridge University Press, 2006), 184.

[19] Moritz A. Drupp et al., "Discounting Disentangled," *American Economic Journal: Economic Policy 2018* 10, no. 4 (November 2018): 109–34.

[20] IPCC WGIII, "Climate Change 2022: Mitigation of Climate Change," 181.

[21] Nordhaus, *The Climate Casino: Risk, Uncertainty, and Economics for a Warming World* (New Haven: Yale University Press, 2013), 188.

[22] IPCC WGIII, "Climate Change 2022: Mitigation of Climate Change," 362.

Distribution

Who will pay for climate damages, and the costs of climate mitigation and adaptation? There are distributional issues within countries — we can't let the burden fall too heavily on the poor. But the larger distributional issue is between countries. 60% of historical GHG emissions come from just three countries: 25% from the US, 22% from the EU, and 12.7% from China.[23] Most developing countries have contributed just a minuscule amount, with India leading the way at 3%. The entire continents of Africa and South America have contributed just 3% each. But a lot of the climate burden falls on these countries in the "global south." "For example, over the past three decades drought in Africa has caused more climate-related mortality than all climate-related events combined from the rest of the world."[24]

Developing countries want to develop their economies to the same level of affluence as the US and Europe. The Paris Agreement relies on the Green Climate Fund, the "Financial Mechanism" of Article 9, which provides a conduit for developed countries to financially support mitigation and adaptation efforts in developing countries. It is supposed to be funded at a level of USD $100 billion per year,[25] but, as of 2020, only about $8.3 billion has been contributed, and only $1 billion of it by the US.[26] Article 8 of the Paris Agreement provides a forum for discussion of "Loss and Damage," i.e. climate damages sustained by developing countries which, in an equitable world, would be compensated by the developed countries who have contributed to the damage. But it was agreed when the Paris

---

[23] Our World in Data, "Who Has Contributed Most to Global CO2 Emissions?," 2019, https://ourworldindata.org/contributed-most-global-co2.

[24] IPCC WGIII, "Climate Change 2022: Mitigation of Climate Change," 1559.

[25] J. Timmons Roberts et al., "Rebooting a Failed Promise of Climate Finance," *Nature Climate Change* 11 (February 18, 2021): 180, https://www.nature.com/articles/s41558-021-00990-2.

[26] Green Climate Fund, "Status of Pledges and Contributions Made to the Green Climate Fund" (Green Climate Fund, July 31, 2020), https://www.greenclimate.fund/sites/default/files/document/status-pledges-irm_1.pdf.

Agreement was signed that Article 8 "does not involve or provide a basis for any liability or compensation."[27]

Another distributional issue is intergenerational: how much of the damages from, and the mitigation and adaptation costs of dealing with, climate change, should be paid by future generations? This factor usually appears in economic analyses as the discount rate, discussed just above.

A final related idea is the just and equitable transition. For example, how do we change our energy systems without the brunt falling on workers, such as those in the oil and gas industry, whose jobs will vanish in the transition? When we stop producing fossil fuels, we must train the workers whose jobs vanish in the transition for new types of work, and provide other types of assistance, such as relocation and counseling.

Climate damages

Climate damages are the monetized value of all the harms resulting from climate change. They will add up to a huge amount of money, and will be much larger if we fail to curb our GHG emissions soon. And the monetary damages estimated by climate economists don't account for all the harms that will occur. As discussed above, many types of harms, such as species loss, harms to human health, and harms to ecosystems are not traded in markets, so economists resort to less-reliable ways of evaluating damages from those harms. Most estimates of climate damages just ignore them.

What we'd like for our high-level economic analysis of climate change is a graph showing global climate damages as a function of the maximum temperature increase: the amount of damages that will occur if we limit heating to 1.5°C, 2.0°C, and so on, up to 8.5°C. These limitations correspond to the Representative Concentration Pathways, which are trajectories of the GHG reductions leading to net-zero emissions later this century. But reality is more complicated than this, and the damages depend on many factors

---

[27] United Nations, "Report of the Conference of the Parties on Its Twenty-First Session, Held in Paris from 30 November to 13 December 2015" (United Nations, January 29, 2016), 8, https://unfccc.int/resource/docs/2015/cop21/eng/10a01.pdf.

beside the temperature increase. The IPCC AR6 reports deal with this complicated reality by formulating a number of Shared Socioeconomic Pathways (**SSPs**), described in Chapter 2 of this book. SSPs are internally consistent sets of assumptions about key drivers, including demography, economic process, technical innovation, governance, lifestyles and relationships among these driving sources. They range from SSP1-1.9, which holds warming to about 1.5°C to SSP5-8.5, which is essentially a business-as-usual fossil-fueled future.

The big-picture framework is fairly clear: annual climate damages, from storms, wildfires, droughts, and other climate-influenced calamities are already high, and will continue to increase as long as temperatures continue to increase, and the temperature will continue to increase as long as we keep burning fossil fuels. Once we stabilize the climate, annual climate damages will stay relatively constant for the next few hundred years, until natural or anthropogenic processes remove GHGs from the atmosphere. That long stream of future damages makes discounting very important in analyzing the trade-offs between mitigation and damages.

There is a lot of divergence in estimates of climate damages, which are generated by modelling programs of various types. The consensus seems to be that climate damages will amount to approximately 2% of GDP for 2°C warming, 6% of GDP for 4°C warming, and 10% of GDP for 6°C warming.[28] But some studies are predicting lower damages, and some are predicting much higher damages: 12% of GDP for 2°C warming, 30% of GDP for 4°C warming, and 35% of GDP for 6°C warming.[29]

One study tallies climate damages from agricultural productivity, undernourishment, heat-related excess mortality, cooling/heating demand, occupational-health costs, hydroelectric generating capacity, thermal power

---

[28] IPCC WGII, "Climate Change 2022: Impacts, Adaptation and Vulnerability," 2497.

[29] Marshall Burke, W. Mathew Davis, and Noah S. Diffenbaugh, "Large Potential Reduction in Economic Damages under UN Mitigation Targets," *Nature* 557, no. 7706 (May 24, 2018): 549–53.

generation capacity, river flooding and coastal inundation.[30] This is of course not a complete list. The study models each type of damages, monetizes and sums the damages, and generates a matrix of predicted GDP-loss graphs, for a variety of SSPs and RCPs. It predicts climate damages may ramp up to 10% of GDP in 2100 if we continue on our fossil-fueled path, but will be one or two percent of GDP if we quickly adopt strict mitigation measures. This study also finds that climate damages will be much higher in Africa and Asia, and South America, than in North America, Europe, and Australia.

We're talking about very high damages. One percent of global GDP is USD $847 billion. We seem to be on a path toward around 4°C warming by 2100, which is likely to ramp up climate damages to 6% of global GDP — around $5 trillion — by the end of the century. Costs will balloon as wildfires, heatwaves, and storms increase, and agricultural productivity declines. Many of the damages will be catastrophic for poor people who can't afford to adapt.

The economic models used to project climate damages don't account well for possible extreme events with low probabilities of occurrence. Possible catastrophic events include the tipping points discussed in Chapter two, such as the disappearance of Arctic sea ice, the collapse of the Greenland Ice Sheet, the Collapse of the Western Antarctic Ice Sheet, a die-back in the Amazon rainforest, the melting of permafrost in cold regions, the breakdown of ocean methane hydrates, and the disruption of ocean circulation patterns. Most of these impacts are irreversible: once we reach the tipping point, the process will continue even if we stop emitting GHGs. Estimates of the probabilities and of the resulting economic damages are uncertain, but one study[31] estimated, based on a poll of experts, that risks from five specific potential tipping points ranged from 0.053%/year/K (where K is the global

---

[30] Jun'ya Takakura and et al., "Dependence of Economic Impacts of Climate Change on Anthropogenically Directed Path," *Nature Climate Change* 9 (October 2019): 737–41, https://doi.org/10.1038/s41558-019--578-6.

[31] Yongyan Cai, Timothy M. Lenton, and Thomas S. Lontzek, "Risk of Multiple Interacting Tipping Points Should Encourage Rapid CO2 Emission Reduction," *Nature Climate Change* 6, no. May 2016 (March 21, 2016): 520–25, https://doi.org/10.1038/NCLIMATE2964.

temperature increase beyond 1°C) to 0.188%/year/K (an 18.8% chance over the next 100 years for each degree of increase in temperature beyond 1°C). For all tipping points, there's a lag time, between 50 and 1,500 years, the time it takes for the full damages to manifest. The study estimated damages from the five specific potential tipping points at between 5% to 15% of GDP. A report from the Swiss Re Institute, the research arm of a large insurance company, multiplies model-based climate damages factors of 5 or 10, to account for "(un)known unknowns."[32] This component of low-probability, high-consequence events in climate damages is sometimes called the "fat tails" problem, referring to a probability graph of damages, where the rightmost part (the tail) doesn't go to zero for large amounts of damages, but stays high enough that it stays "fat." We will mostly ignore these fat tails in the remainder of our economic discussion, but many climate economists think that they are a more important consideration in devising climate-economic policy than the predictable damages. Reducing the likelihood of triggering environmental catastrophes is an important motivation for reducing our GHG emissions now as much as possible.

In conclusion, climate damages—the additional costs imposed upon people and institutions globally, due to global heating—are large now, and will continue to increase annually until we reach net-zero GHG emissions. The damages will become huge. There are virtually no economists who hold that our best economic course is to do no mitigation and just let the damages keep rising. We'll deal below with the question of what would be the economically optimum amount of mitigation, the amount that results in the lowest total of mitigation costs plus damages and adaptation costs.

Adaptation costs

First, some generalities about adaptation. Adaption is a rational economic response of individuals and institutions to climate change. Affected entities

---

[32] Swiss Re Institute, "The Economics of Climate Change: No Action Not an Option" (Swiss Re Institute, April 2021), 10, https://www.swissre.com/institute/research/topics-and-risk-dialogues/climate-and-natural-catastrophe-risk/expertise-publication-economics-of-climate-change.html.

take measures to reduce impacts. For example, the owner of a home on the coast may install a seawall to protect the home from rising seas and storm surges. Or someone living in an area that's getting hotter may install air conditioning. Most adaptation is subject to market forces, and results from self-interested economic decisions by non-governmental entities.

Governments engage in two types of adaptation. First, they adapt for their own operations in the same way as private entities. For instance, they may install air conditioners to improve the working environment of government employees when it makes sense. Second, they can undertake adaptation projects on behalf of their constituents, such as by setting up resilience and cooling centers that individuals can use to cool off during heat waves if they have no air conditioning at home.

The individuals using such cooling centers mostly can't afford to install their own air conditioners, a situation which highlights the fact that many individuals cannot afford the level of adaptation that would restore the quality of their lives to its pre-global-warming state. This is particularly true in poor countries in the Global South.

I will not attempt to quantify adaptation costs in this book. Including adaptation costs in economic models as part of the calculation of climate damages seems to be the current best practice. This makes sense; adaptation is an expected market response to climate damages. But we need to keep the adaptation gap in mind — there may be financial and other barriers preventing an economically optimal adaptation response. Mitigation is more urgent because it will benefit future generations, while adaptation primarily benefits those living now. But it's important for us to financially enable adaptation by the poor and developing countries.

Cost of mitigation

There are many types of actions that can mitigate global heating. Some of them are personal, such as replacing a fossil-fueled car with an electric car. But most of them require organized, public action, such as replacing gas-fired electric power plants with renewable sources of electricity.

The most important mitigation is to stop burning fossil fuels. This must be the centerpiece of our effort to fight global heating. The IPCC WGIII report breaks down GHG emissions by sectors, and an important component of all emitting sectors, with the sole exception of forestry and land use (AFOLU), is "energy systems," i.e. the burning of fossil fuels.[33] They're burned to produce electricity, to propel transport, to heat buildings, and in various industrial processes.

Environmentalists frequently state that switching from fossil fuels to renewable sources for electricity generation will cost nothing, because the cost of new renewables generation has fallen below the cost of new gas-fired plants. But this argument ignores the capital costs of the plants. Currently, about two-thirds of global electricity comes from fossil-fuel power plants, all of which need to be replaced with renewable sources as soon as possible. As discussed above, decommissioning fossil-fuel plants will make them stranded assets, whose value must be written off as expenses by their operating companies. This will be a huge expense. In addition, the electric grid will need massive changes to accommodate the on-and-off nature of renewable generation: solar generation occurs only when the sun is shining, and wind turbines turn only when the wind is blowing. At other times, we need backup capacity to fill in when renewable generation is insufficient. This capacity can eventually come from storage — batteries, for example — that can store power generated when demand is low, and provide the power when renewable sources on the grid are insufficient to meet the demand. This storage will be expensive, as will be the improvements to the power grid required to make the new system work, including building transmission lines for new renewable-energy generation.

The other major area where we need to eliminate fossil fuels is in transport: cars, trucks, boats, airplanes, etc. It costs less to drive a mile propelled by electricity than by gasoline. But replacing all the fossil-fueled conveyances with vehicles powered by renewables will be expensive. And in some cases, like airplanes, we have no good alternative fuel to which we can switch. The best possibility seems to be some renewable biofuel — a plant-based

---

[33] IPCC WGIII, "Climate Change 2022: Mitigation of Climate Change," 246.

fuel that can be grown, refined, and burned, in a renewable cycle without any net CO_2 emissions.

To mitigate at the lowest-possible cost, we should start with the lowest-cost measures. The IPCC AR6 WGIII Report has a chart showing the costs, in USD per ton, for measures that could be taken before 2030 to reduce CO2e emissions.[34] For most of the measures, the costs grow as the reductions increase. For example, the chart shows negative costs for wind and solar energy for reductions up to around 2 GtCO2e/year, with the costs increasing rapidly for further reductions. Several of the measures have costs that are initially negative, meaning that those implementing the measures can save money and reduce GHG emissions at the same time. Examples of measures with initially negative costs include reductions in methane emissions from solid waste and oil and gas production, increasing energy efficiency of lighting, appliances, and equipment, and of aviation, avoiding the demand for energy, increasing the fuel efficiency of vehicles, and switching to e-bikes. Our failure to do these things is a market failure; even though there's a monetary incentive to do them, they're not being done. Most of these measures require investment now to save more money in the future, and not everyone wants to, or can afford to, invest. Governments could incentivize these investments by providing low-interest loans or other financial support.

There are lots of other measures that will cost significant money – up to 200 USD/ton, and there's no market incentive to implement these measures. The IPPC chart lists "carbon sequestration in agriculture" as costing $USD 0-20/ton to start with, with costs rising to between $50 and $100 for higher levels of sequestration. It lists "new buildings with high energy performance" the same way, though costs increase to between $100 and $200/ton as GHG savings from this item reach about 1 Gt/year. A different IPCC chart shows that mitigation costs for keeping temperature increases below 2°C will rise from $80/ton in 2030 to $700/ton by 2100, in keeping with the principle that we will adopt all feasible less-expensive mitigation measures first before moving on to more expensive options.[35] This contrasts

---

[34] IPCC WGIII, 38.
[35] IPCC WGIII, 360

with $9/ton in 2030 increasing to $80/ton in 2100, to keep temperatures below 3°C.

Who will bear the costs of stopping the burning of fossil fuels? Oil and gas companies will lose most of their value on the stock market, because they will be out of business; their oil and gas reserves and infrastructure will lose almost all their value. Auto manufacturers will need to extensively retool their operations to switch from producing gas- and diesel-powered vehicles to electric vehicles. Electricity suppliers will need to abandon their gas- and coal-powered generation plants, and switch to renewables. The stockholders of these corporations, who are, in the end, individuals, will bear the economic costs of these changes. It's difficult to feel sorry for them when climate change has been on the radar screen for decades. Some of their customers will also need to pay more for energy when they switch off fossil fuels. Government policies can affect the distribution of climate costs, of course.

The big picture is that taking aggressive action to reduce GHG emissions as quickly possible will be result in high mitigation costs starting now, and, at the other extreme, doing no mitigation will result in no mitigation costs but will cause high climate damages, increasing gradually but endlessly into the future. From an economics-only viewpoint, we want to reduce the sum of mitigation costs plus climate damages (including adaptation costs). We want to find the optimal point on the curve. It is worth bearing in mind that mitigation costs will eventually be reduced to zero, once we've stopped emitting GHGs, but climate damages will continue, essentially forever.

The IPCC WGIII Report estimates that restricting global average surface temperature increases to 1.5°C will, with medium confidence, result in mitigation costs amounting to a 1.3% to 2.7% GDP reduction in 2050 (and, presumably in many years leading up to 2050).[36] Nordhaus' estimate is in line with this: he calculates that limiting temperature increases to 1.5°C will cost 2.3% of GDP, and achieving a 2.0°C limit would cost 1.4% of GDP.[37]

---

[36] IPCC WGIII, 85.

[37] Nordhaus, *The Climate Casino: Risk, Uncertainty, and Economics for a Warming World*, 177.

But he has two significant caveats: (1) all countries must participate—to achieve these relatively low costs, all countries must take advantage of less-expensive mitigation options in their territories, and (2) the choice of mitigation measures is assumed to be economically efficient—least-cost first. But government regulations currently are a hodgepodge that require some high-cost measures, but not low-cost ones, so our current regulatory path is not economically efficient.[38] And we shouldn't expect all countries to fully participate in climate mitigation.

Total climate costs

To calculate total climate costs, we need to add climate damages to these mitigation costs: 2% of GDP for 2°C warming, 6% of GDP for 4°C warming, and 10% of GDP for 6°C warming.[39] It would make economic sense for us to spend 2% of GDP to limit warming to 2° instead of 4°, since that would save us damages in the amount of 4% of GDP. But we're not on a path to doing this. Given that, in my opinion, the goal of holding warming to 1.5°C is out of reach now, the economic optimum is to hold warming to 2°C, so that mitigation costs and climate damages will each cost around 2% of GDP.

GHG reduction policies

There are two types of measures governments can take to incentivize GHG reductions: command-and-control measures and market-based measures. Adopting either type of measure would send a signal that the world is on a path toward eliminating the burning of fossil fuels by mid-century. As things stand, no one believes this is true, but people would start changing their behavior now if they thought it was true. A command-and-control example is a law prohibiting the burning of fossil fuels without a permit starting in 2050. Market-based measures are the carbon-tax and cap-and-

---

[38] Nordhaus, 177–79.

[39] IPCC WGII, "Climate Change 2022: Impacts, Adaptation and Vulnerability," 2497.

trade. The two market measures are roughly equivalent in economic terms,[40] though they differ in mechanism and politics.

The carbon tax is a tax on GHG emissions. (Note that it's a slight misnomer, as not all GHGs contain carbon.) The amount of the tax would be equivalent to the social cost of carbon (SCC—officially, now, $51/ton officially in the US, or $40-80/ton as estimated by High Level Commission on carbon pricing[41]), which is equivalent to the damages resulting from the emissions. There are two important policy decisions for a carbon tax: at what point in the supply chain is the tax assessed, and what is done with the revenue. For gasoline, using the $51/ton rate for SCC, the tax would cost about 45 cents per gallon. This could be added to the price at the pump, or at the refinery. For fossil power plants it would be simplest for the company generating the electricity to pay, and then pass the costs on to the utilities that distribute the power. My utility, the Los Angeles Department of Water and Power, has an overall GHG intensity of 579 MTCO2e/MWh. The tax for my usage—about 47 kWh/day—would be around 62 cents/day. Industrial users such as cement manufacturers would be required to track their emissions to pay the tax. They'd add it to the cost of their product.

The second big issue for a carbon tax is distribution: who gets the revenue? The tax would probably be regressive in that it would affect poor people more than rich people. But this effect could be neutralized by rebating the tax to those with the lowest incomes, perhaps via their income tax. The entire amount of the tax could be rebated, which would make it revenue-neutral, or a portion of it could be used by the government to fund climate-related research and development, or for some other purpose.

Cap-and-trade is the alternative to a carbon tax. Under a cap-and-trade program, the government establishes annual aggregate limits on each type of GHG emissions (categorized by source: power plants, from cars,

[40] Nordhaus, *The Climate Casino: Risk, Uncertainty, and Economics for a Warming World*, 223.
[41] IPCC WGIII, "Climate Change 2022: Mitigation of Climate Change," 188.

from cement manufacturers) and then sells or gives away allowances for the allowed amount. Each allowance is a mini-permit to discharge a certain amount of GHGs. The allowances may be traded. The distributional issues for cap-and-trade are very similar to those for the carbon tax, if the allowances are sold and generate revenue for the government.

There has been criticism of cap-and-trade by the environmental community. Environmental-justice advocates equate the allowances as permits to continue polluting. Given that many power plants are located in lower-income communities of color, they see this as discriminating against these "front-line" communities. They're primarily concerned about traditional air pollutants such as particulates, NOx and ozone, not CO_2. This criticism is not well taken. Cap-and-trade is a mechanism for lowering emissions, not increasing them. To the degree that cap-and-trade requires the power plant to restrict its GHG emissions, there will be a concomitant reduction in conventional pollutants. And the main purpose of a cap-and-trade system is to reduce GHG emissions, not emissions of conventional pollutants. For this purpose, it doesn't matter where the cuts occur; GHG emissions affect everyone globally. If we can't solve other air-pollution problems via our measures to reduce GHG emissions, so be it; reducing GHG emissions is more urgent.

There are more policies that need to be set correctly for cap-and-trade than for the carbon tax. Which industries does it apply to? How many allowances will be issued for each industry? At which point in the supply chain will allowances be required? What will be done with the revenue? Some cap-and-trade programs in the past were rendered ineffective by the issuance of too many allowances, so there was effectively no cap. But some have succeeded, like California's program.

The main value of the carbon tax and cap-and-trade is that they use the market to incentivize implementation of the lowest-cost GHG limitations first. And they simplify consumers' choices. If the cost to the world of emitting GHGs is factored into the prices consumers and businesses see, they don't have to separately worry about the climate implications of decisions to purchase goods and services. The market will steer them toward the most cost-effective GHG reductions. Command-and-control mechanisms, on the other hand, such as automobile fuel-efficiency standards, frequently mandate GHG reductions that are more expensive than the alternatives. This economic inefficiency slows down our mitigation efforts.

Both types of market mechanisms may be used to limit emissions. Limiting emissions to a particular level—which should gradually decline to zero or almost-zero—is easier under cap-and-trade, because the cap is an emissions limit. As the cap goes toward zero, allowance prices will rise to high levels. Limiting emissions to a certain level with a carbon tax should be possible if regulators monitor emissions and periodically raise the tax level, the price of carbon, to a point where it keeps emissions on the right pathway toward zero.

According to the IPCC AR6 WGIII report, "Carbon pricing incentives may only stimulate incremental improvements, but other instruments may be much more effective for driving deeper innovation and transitions."[42] In other words, economic incentives probably are not sufficient to achieve the level of GHG reduction we need; some command-and-control measures are necessary.

Subsidies are another type of GHG policy. They can be used to spur innovation, when important R&D activities are subsidized. They can also be used to support young industries necessary for our climate response that are still in early stages. But they are currently used mostly to support the status quo. They distort markets, and help keep old, no-longer-viable policies in place. Here are two examples.

Many countries are, perversely, subsidizing fossil fuels. "Removing fossil fuel subsidies would reduce emissions, improve public revenue and macroeconomic performance, and yield other environmental and sustainable development benefits; subsidy removal may have adverse distributional impacts especially on the most economically vulnerable groups which, in some cases can be mitigated by measures such as re-distributing revenue saved, all of which depend on national circumstances;

---

[42] IPCC WGIII, 189.

fossil fuel subsidy removal is projected by various studies to reduce global CO_2 emissions by 1-4%, and GHG emissions by up to 10% by 2030, varying across regions."[43] Subsidies currently supporting fossil-fuel production should be re-directed to support renewable energy research and development.

Similarly, many countries subsidize agriculture and forestry in ways that prevent innovations that would help the climate. The AR6 WGIII Report suggests that these funds should be redirected toward mitigation in the agricultural and forestry sectors.[44] USD $400 billion per year is needed to "deliver the up to 30% of global mitigation effort envisaged in deep mitigation scenarios," and the current subsidies in this area are larger than this amount.

Our current policies attribute GHG emissions to the places where those emissions occur, but the reality of the situation is more complicated. There are, for example, many factories in China that emit GHGs in the process of making products that are consumed in the US. These types of emissions are really caused by demand in the US. We may eventually want to adopt a more nuanced accounting system for GHG emissions that takes the realities of global distribution into account.

Conclusion

The climate crisis is going to cost us, one way or the other. Early and strong mitigation, gradually eliminating fossil fuels from our energy diet, and eliminating deforestation in favor of climate-optimal agricultural and forestry practices, are the most important mitigation measures, but there are many others that will help. Some of them cost nothing, or less than nothing, but we're still not doing them.

The following table shows two options, and the associated costs, expressed as a percentage of global GDP:

---

[43] IPCC WGIII, 46.
[44] IPCC WGIII, 109.

Max Temp Increase	2°C	4°C
Climate Damages	2%	6%
Mitigation Costs	2%	0.5%
Total Annual Damages + Costs	4%	6.5%

These are big numbers: 4% of global GDP is $3.4 trillion, and 6.5% of global GDP is $5.7 trillion. Consider how losing between 4% and 6.5% of your annual income will affect you.

Chapter 5
Law

My law practice: CEQA litigation

I'm a lawyer in Los Angeles. Most of my law practice consists of suing local governments—California cities and counties—under the California Environmental Quality Act (CEQA). My practice illustrates a major defect in environmental law: it's not comprehensive; it's a patchwork of laws each covering a different discrete area, with lots of gaps between the laws.

CEQA requires that, before a state or local government agency approves a project that might have an impact on the environment, the agency must produce a document describing the potential impacts. This document is usually either an Environmental Impact Report (EIR) or a lesser document called a Mitigated Negative Declaration (MND). If some of the impacts are "significant," CEQA requires that they be mitigated to the maximum degree feasible.

CEQA is a very good law. Prior to its enactment in 1970, city councilmembers approving a project wouldn't know whether it hurt the environment. Under CEQA, the city councilmembers are provided with a document laying out the project's environmental impacts, with a determination as to which ones are significant. Unlike National Environmental Policy Act (NEPA) the parallel federal law, CEQA requires that significant impacts be fully mitigated to the extent feasible. If a state or local government entity wants to approve a project that will have significant impacts, even after mitigation, it must adopt a statement of overriding considerations, but it may proceed with the project anyway.

CEQA has been criticized for being used to block projects for reasons unrelated to the environment, and for helping to drive up the cost of housing by adding expensive regulatory hurdles for housing development. But only around 2% of CEQA environmental documents are litigated.[1]

The biggest problem with CEQA is that it addresses the quality of information in the EIR or MND, not the environmental quality of the project. In the Newhall Ranch case I litigated, the developer proposed to build a large mixed-use development in the floodplain of the Santa Clara River, the last free-flowing river in Southern California. What was at issue in the litigation was whether the EIR adequately described the environmental harms: to the river's hydrology, to the plants and animals that would be harmed by channelizing the river, to downstream users harmed by increased salt content in the river water, and by the project's climate impacts. What was, unfortunately, not at issue was whether the project should be built in the river floodplain. That was decided by the County Board of Supervisors, who are subject to political influence by big developers like Newhall. CEQA provides the courts with no power to review such decisions. Courts are allowed to set aside the project approvals and the EIR or MND if those documents don't comply with CEQA's requirements. Then the developer must fix the defects in the environmental document and get the project re-approved.

The thing the court can give a plaintiff in a civil lawsuit is called a "remedy," and my clients seldom want the remedy of setting aside the project approvals and EIR or MND. They generally want physical improvements in the project. One of my clients wants more affordable housing in an apartment project to be built in downtown Los Angeles. Another client wants improved fire-evacuation routes and protections for mountain lions in a project near Santa Clarita. My non-profit, Advocates for the Environment, sues over inadequate analyses of GHG emissions in EIRs and MNDs for development projects, especially warehouses, which are springing up all over Southern California. We want the development projects to be net-zero for GHG emissions.

---

[1] Janet Smith-Heimer, Jessica Hitchcock, and Greg Goodfellow, "CEQA: California's Living Environmental Law" (Berkeley, CA: The Housing Workshop, October 2021), 22, https://rosefdn.org/wp-content/uploads/CEQA-California_s-Living-Environmental-Law-10-25-21.pdf

Project improvements, which the courts can't provide as remedies, can sometimes be obtained in settlement. CEQA lawsuits can be settled before they are filed, during the litigation, or even after the judgment is issued, as long as settlement occurs before judgment is entered. I settle our GHG lawsuits by asking the developers to agree to net-zero projects. In most cases, this means the projects will have no natural-gas hookups, many electric-vehicle (EV) chargers, and solar panels to provide as much on-site electricity as possible. High-quality GHG offsets are then purchased to mitigate any remaining emissions.

In the Newhall case mentioned above, the environmental organizations who had sued Newhall settled after the California Supreme Court held that EIRs for Newhall Ranch violated CEQA's requirements for GHG analysis, and the project would harm an endangered fish, a violation of California's Endangered Species Act. Newhall agreed that the project would be net-zero as to GHGs, and granted other significant concessions in exchange for the organizations' agreement to stop their opposition to the project. That was a good outcome, a big improvement to the project, even though the project is still to be built in the Santa Clara River floodplain.

In a lawsuit against Centennial, another huge project (19,333 housing units, 57,150 residents, 8.4 million square feet of commercial and industrial development on 20 square miles of open-space land), we challenged the EIR's analysis of GHG and fire impacts and won in court. That win gave us enough leverage in settlement that the project became net-zero for GHG emissions, another great outcome.

CEQA detractors would call our actions in these two cases CEQA abuse, because our primary purpose in suing was not to obtain a remedy from the court, but to get the projects improved in a settlement. But this is not abuse; the vast majority of civil cases in general are settled. Settlement is good, because it represents the parties' agreement on an appropriate way to resolve the case. Bringing a case to force a settlement is the normal practice in civil litigation.

From the environmental side, the biggest problem with CEQA is that it governs the environmental documents rather than the project itself. It can prevent the County from approving an EIR that doesn't adequately analyze the Project climate impacts, but it can't prevent Newhall from building its project in the river floodplain. But a good thing about CEQA is that is has very liberal standing requirements: practically anyone can bring a CEQA suit on behalf of the public.

CEQA was patterned after a federal statute, the National Environmental Policy Act (NEPA), enacted by Congress in 1969. Since then, 16 states, New York City, Puerto Rico, and Washington D.C. have enacted laws similar to NEPA. The international-law name for an EIR is an Environmental Impact Assessment (EIA). More than 100 countries now require EIAs to be prepared for projects which may have significant environmental impacts.

Advocates for the Environment's use of litigation to fight climate change by pushing local governments to require net-zero projects flourishes in the legal microclimate of California, but is less effective elsewhere. The main reason comes from the way attorney's fees are settled up at the end of the case. If Advocates for the Environment wins in California, the developer is required to pay our legal fees at our high market rate. If we lose, we are not required to pay the developer's or the local government's fees. This asymmetry is caused by a statute pertaining to lawsuits which achieve a significant public benefit, as ours do. It's different from the normal "American rule" for civil lawsuits, under which each side pays its own lawyers regardless of who wins the case. It's also different from the "English rule," most common in countries other than the US, under which the loser pays the winner's attorneys.

Most environmental litigation is brought by environmental non-profits. There are a few big ones, such as the Sierra Club, with annual budgets around USD $100 million, but most of them are considerably smaller. For them, the risk of paying a developer's attorney's fees if they lose a case is prohibitive. Developers usually hire expensive law firms, and the cost to defend even a simple CEQA case like the ones we bring can run into hundreds of thousands of dollars. And, in many countries, non-governmental organizations (NGOs, mostly non-profits) don't have standing to bring such lawsuits in the public interest; they can be brought only by persons directly affected by the project, such as neighbors.

Marquee climate litigation

My lawsuits against development projects are small potatoes in the overall scheme of things. They're effective because they avoid the discharge of millions of tons of GHGs, but that's still a teeny fraction of what we need to do. Two types of larger-scale litigation attempt to achieve much more. One type has been successful, the other type not so much, at least so far.

Climate lawsuits against governments

Suits against governments attempt to force governments to adopt stricter climate policies. Some of these have succeeded. Here are some examples.

In 2015, a Dutch non-governmental organization (NGO, or non-profit organization), the Urgenda Foundation, along with 900 Dutch citizens, "sued the Dutch government to require it to do more to prevent global climate change. The court in the Hague ordered the Dutch state to limit GHG emissions to 25% below 1990 levels by 2020, finding the government's existing pledge to reduce emissions by 17% insufficient to meet the state's fair contribution toward the UN goal of keeping global temperature increases within two degrees Celsius of pre-industrial conditions. The court concluded that the state has a duty to take climate change mitigation measures due to the 'severity of the consequences of climate change and the great risk of climate change occurring.'"[2] The case was appealed to the Dutch Supreme Court, which affirmed the lower court's ruling.

In 2019 a French NGO, Notre Affaire à Tous (Our Shared Responsibility), along with several other plaintiffs, sued the government of France, alleging inadequate government action on climate change. "On October 14, 2021, the administrative court of Paris ordered the State to take immediate and concrete actions to comply with its commitments on cutting carbon emissions and repair the damages caused by its inaction by December 31, 2022. The court determined that France emitted 62 million extra tonnes of

---

[2] Climate Case Chart, "Urgenda Foundation v. State of the Netherlands," *Climate Case Chart*, 2020, http://climatecasechart.com/non-us-case/urgenda-foundation-v-kingdom-of-the-netherlands/.

emissions from 2015-2018, though it lowered the amount to 15 million tonnes, taking into account the drop in emissions in 2020. The court thus called on France to subtract the emissions caused in excess of its legislative commitments, adding these to the reduction of emissions planned between 2021 and 2022. France has committed to a reduction in greenhouse gas emissions of 40 percent by 2030, compared to 1990 levels, and to reaching carbon neutrality by 2050."[3]

In 2015 a group of young people filed a complaint in federal court in the United States against the US government in the *Juliana* case, alleging that the US' climate policies are violating the plaintiffs' constitutional due-process and equal-protection rights, and failing to protect the atmospheric public trust. The case survived an initial motion to dismiss in Oregon District Court, which set the case for trial in 2019. The US government has gone to incredible lengths to prevent the trial, which would have a stellar cast of witnesses, including climate scientist James Hansen, presenting compelling evidence on how the government's inaction on climate is hurting youth and future generations. There have been numerous petitions to the Ninth Circuit Court of Appeal and the US Supreme Court for stays and writs to stop the trial. The case has been very successful in publicizing the climate crisis, and just getting to trial, which would generate a great deal of public attention, would be an amazing victory. In legal terms, the case has little chance of success because, even if the plaintiffs won in District Court and won the inevitable appeal to the Ninth Circuit, the defendants would ask the US Supreme Court to reverse the ruling, which they would likely do.

Climate lawsuits against oil majors

A number of cities and states have sued the major oil companies—Exxon, Chevron, BP, and so on—for causing climate damages. The most salient claim in these cases is public nuisance, which is defined as "an unreasonable interference with a right common to the general public, such

[3] Climate Case Chart, "Notre Affaire à Tous and Others v. France," *Climate Case Chart*, 2020, http://climatecasechart.com/non-us-case/notre-affaire-a-tous-and-others-v-france/.

as a condition dangerous to health, offensive to community moral standards, or unlawfully obstructing the public in the free use of public property."[4] The idea is that emitting GHGs is a public nuisance because it harms the public's right to a healthy environment free of the impacts of climate change. Oil companies enable the emissions by producing oil and gas that are burned to produce them.

Most of these cases have been going around and around on a game of procedural chutes and ladders orchestrated by the oil companies; as far as I know, none of them has been resolved by a court on the merits. When the cases are filed in state court, the oil-major defendants have them removed to federal court. A federal statute allows such removal so that defendants may have claims brought under federal law heard in federal court. Then the oil companies argue that the federal claims are invalid, and ask the federal court to dismiss the case. The motion to dismiss can be appealed to the federal district court of appeal and then, potentially, to the US Supreme Court. This all can take years, at the end of which the case can be remanded back to be heard in state court. A couple of these cases appear to be on the verge of remand, so they may be eventually decided on the merits in state court.

The distinction between federal and state common law of nuisance adds a complication. As discussed below, much of our law in the US and in countries belonging to the British Commonwealth is common law, meaning judge-made law. When a court in a common-law country issues an opinion resolving a legal case, it may make new law in one of two ways. First, it can just state a principle, such as "no conduct of the defendant, that is specifically authorized by statute, can form the basis for a nuisance claim." Second, it can explain why it is resolving the particular case the way it is. Lawyers in future cases can analogize to the case, based on the explanation, and say their cases should come out the same way (or a different way, if there are significant distinctions). Common law is usually created to fill in the interstices, resolving situations that are not clearly controlled by existing law.

---

[4] "Public Nuisance," in *Black's Law Dictionary* (St. Paul, MN: Thompson Reuters, 2019).

In the *AEP v. Connecticut* case,[5] the US Supreme Court held that the federal Clean Air Act displaces the federal common law of nuisance with respect to CO_2 emissions from power plants. This can probably be generalized to prohibit GHG-related nuisance claims under the federal common law. But state law is distinct from federal law — the two can be completely different on the same subject. There are state nuisance laws which are not displaced or preempted, and state-law nuisance claims based on GHG emissions could still be decided by state courts, if only they could be heard there.

Let's consider a hypothetical nuisance claim, brought by a village in Pakistan that was recently flooded out in a storm whose impacts were greatly increased by climate change. They have $1 million in damage from the storm. The village sues a large emitter, say a California manufacturer of Portland cement. Approximately 4% of global GHG emissions comes from the direct emissions from cement production, and another 3% of GHG emissions come from the indirect use of energy in the production process for a total of 7% of GHG emissions.[6]

A somewhat simplistic allocation of damage would be based on three factors: (1) the manufacturer's share of the global Portland cement market (let's say, 5%); (2) the cement industry's share of global GHG emissions, 7%; and (3) the percentage contribution of climate change to the village's damages. The third factor would need to be estimated with a model of the storm causing the damages. The model would be run twice, to estimate the monetary value of the storm's damage compared to a baseline scenario of the storm with no climate impacts. Let's assume that climate change has made the impacts twice as bad, so that climate change is responsible for 50% of the damage. The village could then seek damages in the amount of $1 million $\times$ 5% $\times$ 7% $\times$ 50% = $1,750. This amount is way too small to justify a lawsuit; the main value of the suit would be its contribution to the public's awareness of climate change.

---

[5] American Electric Power Company, Inc. v. Connecticut, No. 131 S.Ct. 2527 (U.S. Supreme Court June 20, 2011).

[6] Sabbie A. Miller and et al., "Achieving Net Zero Greenhouse Gas Emissions in the Cement Industry via Value Chain Mitigation Strategies," *One Earth* 4, no. 10 (October 22, 2021): 1398–1411, https://doi.org/10.1016/j.oneear.2021.09.011.

Causation is a problem as well. An essential part of a nuisance claim is that the defendant's conduct caused the plaintiff's damage. I learned in my law-school torts class that causation has two parts. The first is called proximate cause. Every one of our actions has impacts that spread out from it, like ripples on a pond after a stone is thrown in. If I break a glass bottle and throw it in the garbage, and a sanitation worker at the waste dump is cut by the glass, there's no proximate cause—the injury is too far away in the causal chain for me to be liable to the sanitation worker. The second part of causation is "but-for": but for the defendant's conduct the plaintiff would not have been injured. This type of causation is missing in our Pakistan-cement hypothetical nuisance claim. The Pakistanis would have been injured even if the cement manufacturer had emitted no GHGs. For most tort claims, but-for causation is required, so the Pakistanis wouldn't have a valid nuisance claim.

I think these causation requirements are outdated. The legal definition of causation should include increasing the risk of harm, and the amount of damages in such cases should be reduced from 100% to a percentage that reflects the amount of extra risk, as we did above when we estimated the damages.

If a court found the cement manufacturer liable to the Pakistanis in our hypothetical nuisance lawsuit, then this same legal theory would make all emitters liable to all those who are being damaged by climate change. I emit GHGs when I drive my gasoline-burning car on a road trip. I am damaging—just a little bit—billions of other people when I do this, and I should be required to pay them. That outcome would be much better achieved by a carbon tax than by litigation. It could result in the same redistribution of money from those who are contributing to global heating to those who are damaged, but would be much more efficient than a million lawsuits.

All of these issues are present in the nuisance suits against the oil majors. And there's one more problem, namely that Exxon, Chevron, and the other oil majors just produce products that other people burn. The oil majors don't directly emit GHGs from their products themselves. This adds yet another link to the chain of causality, and will make it even more difficult to win nuisance suits against the oil majors.

The Dutch nonprofit, Milieudefensie, however, won a case against Shell in the Netherlands in 2021, with the court holding that Shell must reduce its GHG emissions, including emissions from its customers' burning of oil and gas sold by Shell, by 45% below 2019 levels by 2030. The ruling was based on the "unwritten standard of care laid down in Book 6 Section 162 Dutch Civil Code," which requires Shell to observe due care not to harm others, the Paris Agreement, and on EU human rights principles. Shell says it intends to reduce the CO_2 intensity of its oil and gas products, but, since 85% of its emissions come from fuel it sells to customers, it's hard to see how Shell will be able to achieve this level of reduction within 9 years.

Public trust

Professor Joseph Sax's 1970 article on the public-trust doctrine[7] is one of world's most famous law review articles. The doctrine goes back to the 6th-century Code of Justinian and English common law. It requires the government to take special care of public-trust resources that it owns: they may be used only for public purposes, and the government must maintain them for those purposes, and may not sell them.[8] The government acts as a fiduciary, holding these assets in trust for both present and future generations of citizens.

Navigable waters are the resources that were originally protected under the public-trust doctrine. In the *Illinois Central Railroad* case in 1892,[9] the US Supreme Court held that the Illinois legislature did not have the right to sell Chicago waterfront to a private developer because it was subject to the public trust. A few years later the US Supreme Court recognized that the doctrine could apply to wildlife.[10] In the *National Audubon* case,[11] the issue was whether the City of Los Angeles, which had water rights allowing

---

[7] Joseph Sax, "The Public Trust Doctrine in Natural Resource Law: Effective Judicial Intervention," *Mich. L. Rev.* 68 (1970): 471–566,
https://repository.law.umich.edu/mlr/vol68/iss3/3.
[8] Sax, 477.
[9] Illinois Central Railroad Co. v. Illinois, 146 U.S. 387 (U.S. 1892).
[10] Geer v. State of Conn., 161 U.S. 519 (U.S. 1896).
[11] National Audubon Society v. Superior Court, 33 Cal.3d 419 (Cal. 1983).

them to withdraw water from Mono Lake for municipal purposes, could continue such withdrawals even though they were greatly harming the lake. (Water levels had dropped 43 feet.) The Court held that the waters in the lake were subject to the public-trust doctrine, which limited the city's right to withdraw water.

The doctrine could be extended by analogy to the atmosphere, the so-called "atmospheric public trust."[12] Emitting GHGs could be said to violate the public trust by harming the atmosphere, an environmental asset that governments hold in trust for their people. Most US states recognize the public-trust doctrine, as do the governments of India, Philippines, and Kenya.[13] It's not generally available in civil-law countries.

Environmental organizations have filed public-trust lawsuits claiming that GHG emissions are harming the atmosphere in many states in the US. They just need to find a sympathetic judge in a state, such as California, where it would not be too big a stretch to extend the doctrine to cover the atmosphere. Such an extension would be similar to the extension in the *National Audubon* case. As far as I know, none of these lawsuits have been successful yet.

Private enforcement

In 1963 the US Congress enacted the Clean Air Act (CAA), which was the first major environmental statute to include a citizen-suit provision. It's been amended since then to beef up this provision. As it stands now, 42 U.S.C. § 7604 allows "any person" to bring a lawsuit in federal court against any other person who has violated the CAA, or against the US Environmental Protection Agency (EPA), if the EPA has failed to perform a duty required under the CAA. Several other US environmental laws enacted since then also contain citizen-suit provisions, most notably the Clean Water Act.

---

[12] Ken Coghill, Charles Sampford, and Tim Smith, *Fiduciary Duty and the Atmospheric Trust* (New York: Routledge, 2012), 111.

[13] Emma Lees and Jorge E. Viñuales, *The Oxford Handbook of Comparative Environmental Law* (Oxford: Oxford University Press, 2019), 431.

These provisions empower NGOs to enforce important parts of these laws in court, mainly by providing two important things. The first is standing, the right of a particular individual to assert a specific legal claim. Normally, to have standing, a plaintiff must have suffered an actual injury which the court can redress. But NGOs, when they sue under the environmental laws are suing on behalf of the public at large, not to redress injuries to the NGO. The second thing citizen-suit provisions provide to NGOs is a cause of action, a legal claim. The common law provides many causes of action, such as for negligence and breach of contract. But environmental claims must be authorized by statute, and citizen-suit provisions provide the authorization.

Citizen enforcement of environmental laws supplements government enforcement. Government agencies always have too much to do with their limited resources, so they generally have discretion about which laws they enforce in specific cases. Allowing private citizens, especially NGOs, to enforce these laws allows for much more comprehensive enforcement. NGOs are unlikely to bring frivolous enforcement suits because such suits are expensive. They can, in some cases, recover their attorney's fees from the project proponent if they win the suit, but they're always on the hook for their lawyer's fees if they lose.

California has state equivalents of most of the major US federal environmental laws, such as the Clean Water Act, Clean Air Act, and Endangered Species Act, but most of the state statutes do not allow citizen enforcement. One that does is CEQA, the California Environmental Quality Act, the statute requiring environmental impact assessments for major projects conducted or approved by the government. CEQA has relaxed standing requirements; any person who has participated in the administrative process leading to a project approval can sue in the public interest under CEQA.

In the EU, NGOs have much more limited standing to sue; it must be authorized by code. For example, in France the Environmental Code authorizes NGOs that were organized to protect the environment to sue to challenge projects that cause environmental harm to the interests the NGOs were set up to protect. This is much narrower than the standing provided by CEQA. And the attorney's-fees laws are less favorable to NGO plaintiffs in the EU than in the US. The losing party in most lawsuits is required to pay the legal fees of the winning party. Fees tend to be much higher for defendants than plaintiffs because developer defendants usually employ large law firms, which bill at high hourly rates. Potential fee liability for environmental defendants, if they lose the case, is a major deterrent to environmental-enforcement litigation.

China also requires environmental impact assessments for certain major projects, and has started to allow litigation under their EIA law by environmental NGOs. In the "green peafowl case," China's top court affirmed a lower-court ruling halting the construction of a hydropower station in Yunnan province due to an inadequate environmental-impact assessment.[14] But, in general, there has been relatively little climate litigation in China. A law was passed in 2015 that allows certain local NGOs to initiate Environmental Public Interest Litigation, but the Chinese government monitors NGOs carefully and cracks down on those that go too far. According to an expert on Chinese environmental law, "China allows the public-interest litigation, but only to serve the purpose of the Party."[15]

Private enforcement is an important way to increase the enforcement of environmental laws at no cost to the public, but, in many places, business interests don't want a high level of enforcement.

Other climate litigation

Litigation is a lawyer's superpower; it's a power conferred by a license to practice law and the knowledge of how to use litigation to achieve certain ends.

---

[14] Yin Cao, "Biodiversity Ruling to Block Dam Project First of Its Kind in China" (China Daily, February 21, 2022),
https://www.chinadaily.com.cn/a/202202/21/WS62133309a310cdd39bc87ef5.html.
[15] Bloomberg News, "China Wants More Climate Court Cases, But Only the Right Ones" (Bloomberg, June 19, 2021),
https://www.bloomberg.com/news/articles/2021-06-19/climate-litigation-must-navigate-china-s-complex-legal-system.

I, like many other lawyers, have been wracking my brain for several years now to design lawsuits to fight climate change. It's frustrating, because climate change is causing massive harm right now. There's a famous saying in law: "ubi jus ibi remedium." (Lawyers love Latin.) It means "for every wrong there is a remedy." This may be close to true for claims between individuals relating to ordinary life. But it is completely untrue for environmental law.

So we climate litigators are always looking for potentially viable legal claims that might be brought in some legal forum to reduce GHG emissions. Litigation can, under the right circumstances, be extremely effective and efficient in terms of tons of GHG reductions per dollar spent; it can be much more efficient than policy or legislative advocacy, but it requires a lot of creativity and luck. It's deeply frustrating that we can't win a case like *Juliana* against the US government, to force the government to take more action to reduce GHG emissions and meet the goals of the Paris Agreement, or to sue large emitters to make them pay the social cost of carbon for their emissions, which are harming everyone. But there are some cases we can win; we need to creatively expand the legal theories we use in our fight.

Treaties

Treaties are like contracts among States (i.e., countries). Some treaties are bilateral, meaning they are contracts between just two states. Most treaties relating to the environment are multilateral, with many adherents. A good example of the later is the UN Framework Convention on Climate Change, discussed below. It is a contract among the 198 countries that have ratified it.

I'd run several small businesses before I attended law school, and in the course of doing so I negotiated many contracts between my businesses and other entities, such as landlords and customers. When I stepped into my first class at law school, the subject of which was contract law, I thought "finally, I'll get some tips on how these contracts should be written." But, alas, that was not to be. Instead, the contracts course acquainted me with the domestic legal framework that contracts plug into, which controls things like when contracts are made, how they are interpreted, and how they are enforced.

Treaties operate in a similar way, but plug into a different legal framework: that of international law. The Vienna Convention on the Law of Treaties (VCLT) is a multilateral treaty concerning the definition and creation of treaties, and their operation and termination.[16] It has been ratified by 116 countries, but not by the US. In spite of the failure of a few dozen other countries failure to ratify the VCLT, it is generally accepted as binding, customary international law.

The US has an unfortunate habit of failing to sign or ratify international environmental and human-rights treaties, including the VCLT, the United Nations Convention on the Law of the Sea (UNCLOS), the Comprehensive Nuclear-Test-Ban Treaty (CTBT), the Convention on Biological Diversity (CBD), the Convention on the Elimination of All Forms of Discrimination Against Women (CEDAW), and the Rome Statute of the International Criminal Court.[17] The US, as the world's currently most economically and militarily powerful nation, apparently wants to protect its sovereignty and freedom of action, refusing to restrict them through agreements for international cooperation. In my opinion, this course of action is profoundly mistaken. We need a strong international legal order, not only to effectively deal with climate change, but also to deal with other environmental, economic, and human-rights issues. The US, after World War II, was a leader in setting up a strong international order, but has failed to maintain that leadership for some time now.

The three steps to create a treaty are negotiation, signing, and ratification. Multilateral treaties are organized at meetings between the potential state parties. When negotiations have finished, treaties are typically opened for signature. A state's signature usually means that it agrees to pursue whatever measures its domestic laws require in order for it to ratify the treaty. In the US, such ratification requires a two thirds-vote approval by the Senate. Treaties with the EU are ratified by the EU Parliament and

---

[16] Sean D. Murphy, *Principles of International Law* (St. Paul, MN: Thomson/West, 2006), 66.

[17] Anya Wahal, "On International Treaties, the United States Refuses to Play Ball" (Council on Foreign Relations, January 7, 2022), https://www.cfr.org/blog/international-treaties-united-states-refuses-play-ball.

Council.[18] In China, treaties are ratified by the Standing Committee of the National People's Government. Treaties usually specify requirements for them to enter into effect; the requirement is usually that the treaty is signed by a specific number of state parties.

The effect of treaties in domestic law depends on the law of the country. In the US, provisions of a ratified treaty that are "self-executing" are directly enforceable in court, but other provisions may require legislation by Congress to be effective as US law.[19] The same is true in the EU as well, where treaties may require national legislation in the member states to become domestic law there.

Some larger treaties set up a secretariat, an executive institution to administer the treaty, and yearly meetings called Conferences of the Parties (COPs) for the parties to meet to consider issues related to the treaty, such as implementation and amendment. For example, as I write this, the 27th COP (COP27) for the UNFCCC (see below) has just occurred in Egypt, and the 15th COP for the Convention on Biodiversity is about to take place in Montreal.

International dispute resolution

For a contract between private parties, litigation provides a default enforcement mechanism: if one of the parties breaches its obligations under the agreement, the other party can sue it in a court of law, usually for damages arising from the breach. But the parties can opt for a different way to resolve disputes arising under the contract, such as by binding arbitration.

There is no similar default enforcement mechanism for treaties; the text of the treaty specifies how disputes will be resolved. Under some treaties, disputes between state parties may be resolved by the International Court of Justice (ICJ), a United Nations court set up under the UN Charter of 1945.

---

[18] EUR-Lex, "Ratification Process" (European Union), https://eur-lex.europa.eu/EN/legal-content/glossary/ratification-process.html.
[19] Curtis A. Bradley, *International Law in the U.S. Legal System* (Oxford: Oxford University Press, 2013), 41–55.

Other treaties specify arbitration for dispute resolution. Some treaties have much more elaborate dispute-resolution mechanisms. For example, the UN Convention of the Law of the Sea (UNCLOS) establishes its own court, the International Tribunal for the Law of the Sea, which parties may opt to use to resolve disputes. Some treaties, for example the Paris Agreement, have no dispute resolution or enforcement mechanisms at all.

NGOs' participation in dispute resolution under multilateral environmental treaties is unfortunately very limited. In most cases, only State parties can participate.

International climate law – the UNFCCC and the Paris Agreement

In 1994, the United Nations Framework Convention on Climate Change (UNFCCC) went into effect. It's a treaty that has been ratified by 198 countries. It set a goal of stabilizing greenhouse gas concentrations "at a level that would prevent dangerous anthropogenic (human induced) interference with the climate system." It states that "such a level should be achieved within a time-frame sufficient to allow ecosystems to adapt naturally to climate change, to ensure that food production is not threatened, and to enable economic development to proceed in a sustainable manner."[20]

The US Senate ratified the UNFCCC in 1992. This means that the treaty is "the supreme Law of the Land" under Article VI of the US Constitution, giving it the same power as a statute passed by Congress. The UNFCCC contains no provisions that control the behavior of individuals – its requirements apply to state parties only. The US is arguably violating the UNFCCC by failing to do enough to control its climate impacts, and there may be a viable lawsuit against it on this basis, but the treaty cannot be the basis for a lawsuit against a US corporation for its climate impacts because it contains no provisions that restrict emissions by private parties.

---

[20] UNEP, "What Is the United Nations Framework Convention on Climate Change?" (United Nations), https://unfccc.int/process-and-meetings/what-is-the-united-nations-framework-convention-on-climate-change.

In 2015, at the 21st Conference of the Parties to the UNFCCC (COP21), the state parties negotiated the Paris Agreement, a treaty that enhances the implementation of the UNFCCC. It has, as a goal, to "hold[] the increase in the global average temperature to well below 2°C above pre-industrial levels and pursu[e] efforts to limit the temperature increase to 1.5°C above pre-industrial levels."[21] The UNFCCC and the Paris Agreement both recognize that developing countries' obligations under these treaties should be significantly different than developed countries' obligations.

Under the Paris Agreement, countries determine what contributions they should make to achieve the treaty's goals. These are called nationally determined contributions (NDCs). Developed countries should continue taking the lead by undertaking economy-wide absolute emission reduction targets. Developing countries should continue enhancing their mitigation efforts. NDCs should be updated every five years, and should be ambitious efforts that should be improved with each update. Most experts think that the existing NDCs are not sufficient to keep the global temperature increase under 2°C.

At COP15 in 2009, developed countries committed to a goal of mobilizing USD $100 billion per year to finance climate action in developing countries. Reality has fallen somewhat short of this goal, with around $80 billion provided in 2018, 2019 and 2020.[22]

The Paris Agreement is the most important international legal instrument dealing with climate change. Incremental progress is made at the COP each year, but, because any action at the COP requires the consensus of all 196 state parties to the Agreement, any country can block any action, and Saudi Arabia, one of the world's largest oil exporters, frequently uses its veto power to block climate progress. Saudi Arabia also reportedly pressured the IPCC scientists working on the Sixth Assessment Report on Mitigation (the WGIII report) to remove a reference to published literature that found

---

[21] COP21, "Paris Agreement" (United Nations, 2015),
https://unfccc.int/sites/default/files/english_paris_agreement.pdf.
[22] OECD, "Climate Finance and the USD 100 Billion Goal" (OECD, September 22, 2022), https://www.oecd.org/climate-change/finance-usd-100-billion-goal/.

fossil fuels need to be phased out if we're to avoid the worst effects of climate change.[23]

International human rights law

I'm somewhat opposed, philosophically, to the idea of using human-rights law to fight climate change because it seems anthropocentric and speciesist to me. What about the rights of non-human animals? Don't they have any rights? There is a growing movement espousing the rights of nature, and of non-human animals, but it hasn't yet reached the point where such rights can be enforced. But when I'm being practical, I see that this relatively new body of human-rights law might help the cause.

Rights

A right is "a power, privilege, or immunity secured to a person by law."[24] The idea of rights under the law goes back thousands of years in Western culture, at least back to the Code of Justinian in the early Middle Ages. Theoretically, the violation of one of my rights gives me a cause of action against the violator under the law, and the law will provide a legal remedy for the violation. Rights can protect me from harmful actions by other individuals. For example, if someone walks up and punches me in the face out of the blue, that person has violated my right to be free of physical violence and I can sue the person for assault and battery.

One of the most important functions of legal rights is to protect persons from wrongful acts by their government. They are especially important when they protect minorities in a democracy from unfair actions by the majority. These types of rights are usually found in State constitutions or other high-level organizing instruments, and can be used to invalidate statutes passed by the legislature when they conflict with basic rights.

---

[23] Greenpeace International, "Saudi Arabian Negotiators Move to Cripple COP26 – Greenpeace Response" (Greenpeace International, November 7, 2021), https://www.greenpeace.org/international/press-release/50547/cop26-saudi-arabia-negotiators-cripple/.

[24] "Right," in *Black's Law Dictionary* (St. Paul, MN: West Publishing Co., 2019).

Rights can be positive or negative. For example, the First Amendment to the US Constitution states that "Congress shall make no law ... abridging the freedom of speech." This right to free speech is a negative right: it prohibits the government from restricting what one can say. If it were a positive right it would say that every person has the right to express themselves as they wish. Though it was enacted as part of the Bill of Rights in 1791, it applied only to the federal government until 1868, when the Fourteenth Amendment made it applicable to state governments in the US as well. One consequence of its being a negative right is that it doesn't restrict the actions of private persons or corporations. It doesn't prohibit your employer, for example, from requiring you, in an employment contract, not to make any public statements on a particular issue, unless your employer is a government agency. But the right is "hard law," and directly enforceable. The US is the only country where the constitution has been directly enforceable as hard law since the beginning.[25] If a governmental entity in the US tries to restrict what I can say, I can bring a lawsuit in court to enforce my right to say what I want. If a legislative body passes a law restricting what I can say, I can bring a lawsuit and ask the court to invalidate the law as being unconstitutional.

Rights in most other countries, including the EU, tend to be soft-law positive rights. In France, for example, the preamble to the current constitution invokes the 1789 Declaration of the Rights of Man, and Article 11 of that declaration states that "the free communication of thoughts and opinions is one of the most precious rights of man: any citizen can therefore speak, write, and print freely, except to answer for the abuse of this freedom in cases determined by law." This is a positive constitutional right, but it's less absolute than in the US. The last clause means that, unlike in the US, the legislature is free to restrict freedom of speech.

Human rights

Human rights law came into its own after World War II. The UN Charter, which entered into force in 1945, established the United Nations and, under

---

[25] Michel Rosenfeld and András Sajó, *The Oxford Handbook of Comparative Constitutional Law* (Oxford: Oxford University Press, 2012), 112.

it, the International Court of Justice. The UN General Assembly adopted the Universal Declaration of Human Rights in 1948. It was soon followed by the International Covenant on Economic, Social and Cultural Rights and the International Covenant on Civil and Political Rights. These three instruments together constitute what is sometimes called the "International Bill of Rights." The declarations are technically soft law, in that they are not justiciable—they don't provide a legal cause of action when they are violated. But they may still be enforceable in some circumstances as customary international law.[26] They are almost universally acknowledged to be statements of rights possessed by all humans.

Regional agreements confirm and extend the Universal Declaration: the European Convention on Human rights (1950), the American Convention on Human Rights (1969), and the African Charter on Human and Peoples' Rights (1981).

Right to a clean and healthy environment.

On July 28, 2022, the UN General Assembly adopted a right to a sustainable and healthy environment.[27] Such a right is contained in 100 national constitutions.[28] A constitutional right to a clean and healthy environment is therefore fairly widespread, but it is almost always soft law, meaning it is not justiciable and can't be directly enforced in court. Why is this?

Legal rules operate at various levels of generality. Legislators, and judges who make new law, are wary of general rules because they may have unintended consequences. I represented the Sierra Club a few years ago in litigation against the California Coastal Commission. In its equivalent to an

---

[26] Rhona K.M. Smith, *International Human Rights Law,* 10th ed. (Oxford: Oxford University Press, 2022), 60.

[27] UN News, "UN General Assembly Declares Access to Clean and Healthy Environment a Universal Human Right" (New York: United Nations, July 28, 2022), https://news.un.org/en/story/2022/07/1123482.

[28] John Knox, "Constructing the Human Right to a Healthy Environment" (Annual Review of Law and Social Science, July 27, 2020), https://doi.org/10.1146/annurev-lawsocsci-031720-07-4856.

environmental impact report, the Commission had evaluated only the coast-related impacts of a project, even though it was the only agency reviewing the project. I argued that the Commission needed to evaluate all of the project's impacts, not just the coastal-related impacts. The California Court of Appeal ruled in the Sierra Club's favor, but the ruling was based on other reasons, ones we barely discussed in the briefing or argument. My opinion is that the court was too nervous about issuing an opinion requiring the Coastal Commission, along with twenty other state agencies that use the same type of process, to evaluate all of a project's impacts when it is the only agency evaluating the project because of fear of unintended consequences. It would have been the right ruling in this case, and the right law in general, but the court wouldn't go there, out of timidity or caution. It would have been better for them to take the leap of faith and issue a broader ruling, because it would be the right law.

How could a constitutional right to a clean and healthy environment be made justiciable in a way that wouldn't cause a lot of collateral damage? The extreme case would be allowing anyone to sue anyone who violated the right, so I could sue my neighbor for burning wood in his fireplace because the combustion releases GHGs and therefore harms the climate, which is part of my environment. If enacted this way, everyone on the world would have a claim against half the people in the world, quadrillions of potential lawsuits. That is obviously impractical.

One possible adjustment would be to make it a negative right against the government, like free speech in the US. People would have a claim against their government if it violated their right to a clean and healthy environment. A number of the European human-rights cases like *Urgenda* were won partially on human-rights grounds like this. But the right fits awkwardly into this context. Governments need to make policy trade-offs, and we can't adopt a rule requiring an absolutely pristine environment. Such a policy would kill most of our economic activity. A policy that any activity must halt if it harms the environment is impractical for the reasons discussed in the previous paragraph.

These hypotheticals illustrate another problem with the right: it's difficult to make it absolute. Most human rights are absolute. I have an absolute right not to be tortured. I can vindicate that right in court (at least against the government). But my right to a clean and healthy environment, if it is an absolute right, is violated many times each day. The air I breath is polluted, from power plants, from vehicles, from industrial facilities and fireplaces, and wildfires. Allowing me to sue all these emitters because they're violating my right to a clean environment is impractical and inefficient.

In 2020, the UN Special Rapporteur on the issue of human rights obligations relating to the enjoyment of a safe, clean, healthy, and sustainable environment issued a report on good practices relating to the right to a healthy environment.[29] It summarizes many best practices in national laws dealing with the environment, but doesn't suggest a practical way to make such a right justiciable. It does advocate for access to justice and lists many laws that should be adopted by states for environmental protection. But the enforcement it envisions would allow citizens to sue the government for failure to uphold such laws, and not to directly enforce the laws. That's the most practical way to interpret the right in general.

But, even if they are not directly justiciable by individuals, such constitutional and international rights to a clean and healthy environment are valuable because they increase awareness of the environment, and can provide interpretative guidance to judges when deciding cases based on domestic environmental laws.

National legal systems

I want to discuss climate laws in various countries, but some background on their legal systems will be helpful first. In this section I focus on legislation, not litigation.

---

[29] UN Human Rights Council, "Right to a Healthy Environment: Good Practices" (United Nations Human Rights Council, 2020 209AD), https://documents-dds-ny.un.org/doc/UNDOC/GEN/G19/355/14/PDF/G1935514.pdf?OpenElement.

Common law vs. civil law

In England, following the Norman conquest in 1066, the law was made by judges' decisions, which created precedents, which accreted into a comprehensive body of law. In common-law systems like England's, there is no systemized code specifying the law; to find the law one must hunt through prior court decisions. Several hundred years later, the English Parliament acquired the legislative power, the power to enact statutes, which could override judge-made rules. The body of judge-made law was called the "common law" because Henry II, in 1154, created a unified "common" system of courts throughout the country. Countries that were British colonies, such as the US, Canada, and Australia, are governed by the common law. Common-law systems tend to be adversarial, with lawyers for the two contending parties arguing their respective cases before a neutral judge. Common-law systems tend to have constitutions containing negative, hard-law rights.

The other important system is called "civil law." In countries with a civil-law system—most of the non-common-law countries—a written code of law, enacted as a statute or series of statutes, is primary. The civil-law tradition comes from Roman law, and especially the Code of Justinian, adopted in the 6th century in the Roman Empire. The judge, in a civil-law system, is supposed to apply the code to the facts at hand; judicial precedent is much less important than in the common-law system. Civil-law cases tend to be less adversarial; the judge is sometimes responsible for investigating the case, in addition to deciding it. Civil-law countries tend to have soft-law rights that are not directly justiciable.

In practice, the difference between the two systems is less than one might think. Legislatures enact statutes in common-law countries; those statutes override judicial precedent, and must be followed by the courts. The statutes are codified, which means they are integrated into a fairly comprehensive code that includes most of the applicable statutory law. In the US, for example, there is the Code of Laws of the United States of America (U.S. Code or USC). It contains most civil and criminal laws arranged in a logical fashion. For example, the Clean Air Act, a federal statute, is codified at 42 U.S.C. §§ 7401 to 7671q—volume 42, sections 7401 to 7671q. And, in California, the statutory law is codified in a number of named codes, such as the Government Code, the Public Resources Code, and the Penal Code. This is not very different from France, a civil-law country, which has many similar codes, including the penal code, environment code, and electoral code. And courts in civil-law countries sometimes cite prior court decisions in their written opinions; precedent has some weight, even in these countries.

Environmental Law is mostly statutory, even in common-law countries. At the federal level in the US, there are a dozen or so major environmental statutes, such as the Clean Water Act, the Clean Air Act, the National Environmental Policy Act, and the Endangered Species Act. Most states in the US have enacted their own, similar, environmental laws. The fact that environmental law is mostly statutory diminishes further the differences between common-law and civil-law countries, for environmental purposes. In both systems, judges' primary role is to interpret the text of the statutes, but courts also fill in the legal interstices. Legislatures can't think of every eventuality and complication when they enact a law; there will always be special circumstances not envisioned by the drafters of laws, and courts must deal with those situations by crafting new rules that fill the gap and harmonize with the rest of the laws. They do this in civil-law as well as common-law countries.

Types of domestic law

Constitutions: These days, most countries have written constitutions, the obvious exception being Great Britain. The US Constitution of 1789 is the oldest written, codified constitution. It superseded the Articles of Confederation, which set up a federal government that was much too weak to effectively govern the country. A constitution provides a legal framework for a country, and is usually the highest law of the land.

Statutes: Constitutions usually provide for a legislative body, like the US Congress or the French Parlement. Those bodies enact legislation in the form of statutes, which are laws binding on those residing in the country. The statutes are often integrated into a code, which is an organized compendium of most laws currently in effect in the country. Statutes must usually be consistent with the constitution, and courts may be empowered to nullify them if they are not.

Regulations: Legislatures frequently delegate the details of administering a statute they've enacted to a governmental agency. The statute authorizes the agency to enact regulations, which have the force of law. Other laws — e.g. the Administrative Procedures Act in the US — govern the agencies' process in adopting these regulations. They must be consistent with the enabling statute and, of course, with the constitution.

Levels of Government: Countries all have centralized national governments, with national legislatures. Some countries are divided into states or provinces, which may have legislatures. And local governments — cities and counties — often have legislative bodies, city councils and county boards of supervisors, for example. Usually, provincial legislation must be consistent with any related national legislation, and the scope of provincial legislation is limited to the provincial territory. The scope of local legislation is usually much more limited: cities and counties can make laws only on certain matters that predominantly affect local interests; chief among these is land use.

United States

The US Constitution has been in effect since 1789, over 230 years. This is a very long time for a constitution to remain essentially unchanged, especially when the constitutions' lifetimes have averaged about 19 years.[30] France has had — depending on how one counts — around 14 constitutions in that time, starting with the 1791 constitution establishing a limited monarchy, ending with the 1958 Constitution,[31] which is still in effect, though it has been amended 24 times since it was adopted.

---

[30] Zachary Elkins, Tom Ginsburg, and James Melton, *Conceptualizing Constitutions: The Endurance of National Constitutions* (Cambridge, UK: Cambridge University Press, 2009), 36–64.

[31] Gilles Thevenon, *Histoire des Constitutions: Vie politique française 1789 à 1958* (Lyon: Chronique Sociale, 2017), 177.

The US Constitution is much harder to amend than the French Constitution. In the US, an amendment must be adopted by a two-thirds vote of both houses of Congress, then ratified by three-quarters of state legislatures. (There are alternatives which have not been used thus far: the amendment could be adopted by a constitutional convention convoked by the legislatures of two-thirds of the states; and an amendment could be approved by state ratifying conventions in three-quarters of the states, if Congress chooses this over ratification by state legislatures.) In France, a constitutional amendment may be adopted by a simple majority vote of the two houses of the legislature, then approved by popular vote in a referendum. In the alternative, if the amendment was proposed by the President, the amendment may be approved by a three-fifths vote of a Congress composed of the members of both legislative chambers, without a popular vote.

The big problems we're having with the US Constitution are partly caused by the difficulty of amending it. The last amendment took effect 30 years ago, in 1992. It delays laws affecting Congressional salaries from taking effect until after the next election of representatives. It was proposed in 1789, and took 202 years to be ratified. The preceding, 26th, amendment prohibits the denial of the right of US citizens, 18 years of age or older, to vote on account of age. It was ratified in 1971, over 40 years ago. In 1972, Congress approved the Equal Rights Amendment, which would guarantee gender equality. It has the support of about 80% of likely voters in the US, but has been in legislative limbo for decades since it hasn't been approved by the required number of state legislatures.

Here is a partial list of things which need to be fixed in the US Constitution:

- Equal rights for women (above);
- Right to an abortion;
- Right to gay marriage and marriage between persons of different races;
- Right to privacy;
- Right to a clean and healthy environment;
- Senate is undemocratic – a Senator from Wyoming represents around 300,000 people (half the state's population), while one from California represents about 20,000,000 people;

- Senate is undemocratic – the filibuster rule allows 40 senators, representing a small minority of the population, to block legislation desired by the majority;
- Electoral college system for Presidential elections is undemocratic— it should be replaced with a nationwide popular vote;
- Gun rights should be restricted; polls show most Americans are in favor of more restrictions on owning and carrying firearms;
- Separation of church and state is insufficient now; it should be clear that religious symbolism and speech should be prohibited in government facilities and functions;
- Corporations' donations to political campaigns should be subject to limits set by the legislature, and not treated as free speech protected under the First Amendment.

I believe that all of the above changes are supported by a majority of Americans; it is undemocratic that we cannot enact changes in the law that are supported by the majority. There is a similar list of important changes in statutory law that are supported by the majority, but which Congress will not pass, due to the political gridlock. A good example is the scope of the "waters of the United States" that are protected under the Clean Water Act. The Supreme Court interpreted it in a difficult-to-apply way in the *Rapanos* decision in 2006, and the definition has been unsettled ever since; it's unclear whether the CWA protects wetlands adjacent to tributaries, or intermittent and impermanent streams. Property owners don't want the protections to apply to their land because the CWA requires federal permits for construction within protected waters. Environmentalists want broad protections for those waters. Congress could solve the problem by amending the CWA with a clear definition, but passing such a law is politically infeasible, so the definition rests with the US Supreme Court, purporting to interpret the text of the law.

Similarly, when the US Constitution can't be changed, as a practical matter, then it's very important how it is interpreted. Indeed, US Supreme Court decisions over the last century have filled in, to take the place of constitutional amendments. For example, the Fifth Amendment says: "No person shall be...deprived of life, liberty, or property, without due process of law." The section of my Constitutional Law textbook concerned with the Supreme Court's interpretation of just the due-process portion of this clause, split into "Procedural Due Process" and "Substantive Due Process" occupies 248 pages. And it is just a summary, leaving out many details. In detailing out what this clause means, the Supreme Court is essentially legislating. The Court's approach to this type of interpretation has changed a lot over the last hundred years. The Warren Court in the 1950s and 1960s felt that the Constitution should keep up with the times. It read a right to privacy into the Constitution in the *Griswold* case. It ended segregation in schools in the *Brown v. Board of Education* case. It required those arrested to be given *Miranda* warnings. And, in 1973, after Justice Warren retired, the Court found a right to an abortion, in *Roe v. Wade*.

The current Supreme Court is revisiting and overturning landmark decisions like this. An influential minority of the justices is taking an "originalist" approach to interpreting the Constitution, asking what the framers had in mind when they wrote it. The originalists criticize the Warren court's interpretations as not being interpretations at all—they feel the Court just made up the rule it thinks should apply, purporting to base it on a very expansive interpretation of the Due Process Clause that is untethered to the Constitutional text. The originalists prefer to use the constitutional text, and the framers' intent, as the touchstone for interpretation, because it connects the interpretation to the text in a much more concrete way. But does it make sense, when considering whether there should be a right to an abortion, to base it on what James Madison would have thought about the issue? Or what he would have thought about the Court reading new rights, such as the right to privacy, into the Fifth Amendment? Things have changed a great deal in the last 230 years, and we need to keep the Constitution up-to-date. It would be best if we could amend it to conform to our current needs. But, since we can't do that now, because of US politics, we need to update it by interpreting it in line with our current culture and needs, using the Warren-court approach instead of the originalist approach.

The US has a federal system, with the 50 states retaining their sovereignty. The US Constitution establishes a federal government with legislative powers limited to those powers listed in Article I, section 8. Federal law enacted under one of the enumerated legislative powers preempts conflicting state law.

The US' latest Nationally Determined Contribution (NDC) under the Paris Agreement, submitted to the UNFCC in April 2021, calls for reducing the US' GHG emissions by 50-52 percent below 2005 levels by 2030.[32] This includes a goal of reaching 100 percent carbon pollution-free electricity by 2035. The NDC estimates US 2005 emissions as 6.6 Gt CO_2e, which means the US' plans to reduce its emissions to 3.3 Gt by 2030. US emissions in 2020 were 5.2 Gt. Reducing to 3.3 Gt by 2030 would require a 4.4% reduction each year until then. This seems possible, given that the US reduced its GHG emissions by 11% in the 2019-2020 year.[33] But it is very unlikely, in my opinion, that we'll have a carbon-free electricity sector by 2035; thirteen years is not enough time to replace all the coal- and gas-fired power plants with renewables, let alone to add enough storage and change the distribution system to accommodate 100% renewable power.

The US doesn't have a comprehensive climate law. In 2015, the Obama administration promulgated regulations known as the Clean Power Plan that would have substantially reduced emissions from the electric power sector. It was successfully challenged in court, and then withdrawn by the Trump administration.

In 2022, Congress passed the Inflation Reduction Act (IRA). One commenter characterized it as "all carrot, no stick" It provides a great deal of money for a wide variety of programs that will help reduce the country's GHG emissions, including:

- Expanding and extending tax credits for clean energy;
- Investing in manufacturing solar panels, batteries and other clean energy technologies in the US;

---

[32] "The United States of America Nationally Determined Contribution," April 2021, https://unfccc.int/sites/default/files/NDC/2022-06/United%20States%20NDC%20April%2021%202021%20Final.pdf.

[33] U.S. EPA, "Inventory of U.S. Greenhouse Gas Emissions and Sinks," April 4, 2022, https://www.epa.gov/ghgemissions/inventory-us-greenhouse-gas-emissions-and-sinks.

- Funding for low-income families to electrify their homes;
- Removing barriers to community solar;
- $3 billion for the US Postal Service to electrify its fleet;
- $1 billion for clean school and transit buses and garbage trucks;
- $3 billion to clean up pollution at ports by installing zero-emissions equipment and technology;
- Tax credits for electric vehicles;
- $3 billion for community-led projects in areas experiencing the disproportionate impacts of pollution and climate change;
- $20 billion to help farmers and ranchers shift to sustainable practices like crop rotation and cover crops;
- $2.6 billion in coastal resilience grants;
- $1 billion to ensure federal agencies can conduct robust environmental review on large projects using federal funds on federal lands;
- Tax credits for CCS at coal plants.[34]

One study models the reductions we can expect from the IRA, concluding that the IRA will reduce US GHG emissions to around 4 Gt by 2030.[35] This gets us a bit more than halfway toward where we need to be.

Shortly after Barack Obama became President in 2008, the House of Representatives passed the American Clean Energy and Security Act of 2009 by a vote of 219 to 212. It was written by Reps. Henry Waxman and Edward Markey and would have established an economy-wide greenhouse gas cap-and-trade system.[36] The bill never passed the Senate, so it never became law. There are two cap-and-trade systems operated by

---

[34] Earthjustice, "What the Inflation Reduction Act Means for Climate," August 16, 2022,
https://earthjustice.org/brief/2022/what-the-inflation-reduction-act-means-for-climate.
[35] Princeton University Zero Lab, "Preliminary Report: The Climate and Energy Impacts of the Inflation Reduction Act of 2022," August 2022,
https://repeatproject.org/docs/REPEAT_IRA_Prelminary_Report_2022-08-04.pdf.
[36] "Congress Climate History" (Center for Climate and Energy Solutions),
https://www.c2es.org/content/congress-climate-history/.

US states, though. Nine eastern states jointly set up a regional cap-and-trade program known as the Regional Greenhouse Gas Initiative, which applies to CO_2 emissions from fossil fuel-fired power plants with 25 MW or greater capacity. An emissions cap gradually declines, which increases the cost of the allowances that are required for the covered power plants to emit CO_2. California operates a similar system, which applies to large emitters (more than 25,000 metric tons of CO_2e), including power plants, industrial facilities, and fuel distributors.[37]

European Union

The European Union (EU) has become a federated democracy like the United States, but it's taken decades for it to evolve from modest beginnings in the Treaty of Paris of 1951. That treaty established the Coal and Steel Community in Western Europe, setting up coal-and-steel cooperation among the state signatories, France, Italy, West Germany, Belgium, Luxembourg, and the Netherlands.[38] The Treaty of Rome, in 1957, established the European Community (EC, originally called the European Economic Community) and created a common market based on the free movement of goods, people, services and capital.[39] It also coordinated customs, agricultural, trade and transport policies. It established some of the important EU institutions, such as the Council of Ministers, the European Commission, the European Parliament, and the Court of Justice.

The two main treaties that act as a quasi-constitution for the EU are the Treaty on European Union (TEU), negotiated in Maastricht in 1992[40] and

---

[37] Seth Kerschner, "United States: Greenhous Gas Emissions Trading Schemes" (Lexology/White & Case LLP, September 20, 2017),
https://www.lexology.com/library/detail.aspx?g=0f6bf054-27dd-4cc0-b856-107b1ad0854e.
[38] "Treaty Establishing the European Coal and Steel Community" (1951),
https://eur-lex.europa.eu/legal-content/EN/LSU/?uri=CELEX:11951K/TXT.
[39] "Treaty Establishing the European Economic Community" (1957),
https://eur-lex.europa.eu/EN/legal-content/summary/treaty-of-rome-eec.html.
[40] "Treaty on European Union" (1992),

the Treaty on the Functioning of the European Union (TFEU), which went into effect in 2009.[41] The treaties have been amended many times as more member-states joined the EU and as the scope of the federation enlarged.[42] In these treaties, the 27 current member nations agree to cooperate in many areas, and to enact national laws that are consistent with certain EU policies. The treaties set up a number of important institutions. The European Council is a high-level body composed of the heads of member states and the President of the Commission. It is concerned mostly with political and policy issues. The Council of the European Union (why did they have to create two councils with similar names?) is the EU's major legislative body; it also shares executive power with the Commission. The Commission of the European Communities has 20 members, which are mostly appointed by member states. It manages a large bureaucracy that makes up most of the EU's executive branch. The European Parliament is primarily a legislative body, though it also has a consultative role, and the right of approval of many acts by the Commission.[43] Members of the Parliament are elected in a direct popular vote in the member countries.

Though the legal structures are completely different, the EU's federal system is in many practical ways similar to that of the US. The 50 states in the United States retain a lot of their sovereignty, just as the 27 EU member states do. In both cases there are complex rules governing the relationship between legislation passed by Congress or the EU's institutions and legislation passed by the states or member states. The power of the federal/EU legislatures to pass legislation is restricted to certain areas, the enumerated powers of Congress granted by the US Constitution, and the powers granted to the EU institutions by the TEU and TFEU (and other

---

https://eur-lex.europa.eu/legal-content/EN/TXT/PDF/?uri=CELEX:11992M/TXT&from=EN.

[41] "Treaty on the Functioning of the European Union" (2007), https://eur-lex.europa.eu/EN/legal-content/summary/treaty-on-the-functioning-of-the-european-union.html.

[42] "Chronological Overview" (EUR-Lex, 2020), https://eur-lex.europa.eu/collection/eu-law/treaties/treaties-overview.html.

[43] Philip Raworth, *Introduction to the Legal System of the European Union* (Dobbs Ferry: Oceana Publications, 2001), 76–81.

treaties). The rules also specify how and whether the federal government can compel states or member states to pass and enforce legislation, and how and whether legislation passed by the federal government takes effect in states and member states.

The EU enacted the European Climate Law in 2021.[44] It aligns with the EU's 2020 NDC, committing to reducing GHG emissions by 55% below 1990 levels by 2030.[45] It also sets a goal of climate neutrality by 2050. It was enacted as a regulation, and states that it is "directly applicable in all Member States," meaning that it has the force of law in the member states, potentially overriding any conflicting national legislation. It requires the Commission to monitor, and report every five years on the EU's and member states' progress toward meeting these goals, and provides the Commission with very limited enforcement powers, should a member state fail to implement policies consistent with the European Climate Law.

The EU has also set up a cap-and-trade program, called the Emissions Trading System (ETS). It covers CO_2 emissions from electricity and heat generation, commercial aviation within the European Economic Area, and energy-intensive industry sectors including oil refineries, steel works, and production of iron, aluminum, metals, cement, lime, glass, ceramics, pulp, paper, cardboard, acids and bulk organic chemicals.[46] It seems to have been effective: one study estimated that it reduced carbon emissions in Europe by between 8.1% and 11.5%.[47]

---

[44] "European Climate Law" (European Commission, 2021), https://climate.ec.europa.eu/eu-action/european-green-deal/european-climate-law_en.

[45] European Commission, "Update of the NDC of the European Union and Its Member States," December 17, 2020, https://unfccc.int/sites/default/files/NDC/2022-06/EU_NDC_Submission_December%202020.pdf.

[46] European Commission, "EU Emissions Trading System (EU ETS)," https://climate.ec.europa.eu/eu-action/eu-emissions-trading-system-eu-ets_en.

[47] Patrick Bayer and Michaël Aklin, "The European Union Emissions Trading System Reduce CO2 Emissions despite Low Prices," *PNAS* 117, no. 16 (April 6, 2020): 8804–12, https://doi.org/10.1073/pnas.191812811.

China

China is the world's largest GHG emitter, though its per-capita emissions are a fraction of the emissions of the US. It is key for controlling climate change. China is in transition from a developing to a developed country. In its NDC,[48] China commits to controlling its CO_2 emissions so they peak before 2030 and achieving carbon neutrality by 2060. This falls significantly behind most of the other NDCs, which commit to reaching net-zero by 2050.

China needs more energy as it builds out its infrastructure, and it has been rapidly building coal-fired power plants. Coal's share in the country's primary energy consumption has been reduced from 69.2% in 2010 to 56.8% in 2020;[49] this is still very high, especially since China continues to build coal-fired power plants at a high rate. In 2021, China started projects that result in 33 gigawatts of coal-based power generation, three times more than the rest of the world combined.[50]

China has no climate law per se. In 2018, China put the newly established Ministry of Ecology and Environment in charge of responding to climate change. It has enacted a large number of policies aimed at integrating the country's response to climate change into its development plans.

China has set up seven pilot cap-and-trade systems in various parts of the country, covering electric power and a variety of other sectors and, based

---

[48] "China's Achievements, New Goals, and New Measures for Nationally Determined Contributions" (China, 2021), https://unfccc.int/sites/default/files/NDC/2022-06/China's%20Achievements%2C%20New%20Goals%20and%20New%20Measures%20for%20Nationally%20Determined%20Contributions.pdf.

[49] "How Is China Tackling Climate Change?" (Grantham Research Institute on Climate Change and the Environment, 97/25/22), https://www.lse.ac.uk/granthaminstitute/explainers/how-is-china-tackling-climate-change/."Responding to Climate Change: China's Policies and Actions" (State Council Information Office of the People's Republic of China, October 27, 2021), http://www.scio.gov.cn/zfbps/32832/Document/1715506/1715506.htm.

[50] Mark Green, "Chinese Coal-Based Power Plants" (Wilson Center), https://www.wilsoncenter.org/blog-post/chinese-coal-based-power-plants.

on the results of these, launched a national carbon emission trading market in 2021.[51]

China is the most important country for controlling climate change. Diplomacy by the US and other important countries seems to be the most likely path to convince them to reduce emissions as quickly as possible.

An international constitution

Scholars have relatively recently started talking about an international constitution, particularly in the area of human rights.[52] One function of constitutions is to limit the exercise of public power, and the human-rights treaty regime does that by providing rights that cannot be abridged by government to the residents of countries that have ratified the treaties.

Conclusion

In this chapter I've discussed climate litigation and climate laws in representative countries. Litigation is the most significant way NGOs can help fight global heating. The ability of litigation to work in a particular local context depends on what one might call the legal microclimate. A direct attack on the national government, as in Urgenda, asking the court to order the government to further reduce the country's GHG emissions, worked in the Netherlands but would not work in the US, partly due to the current composition of the US Supreme Court. But the kind of Environmental Impact Assessment lawsuits I do in California depend on unique features of the California legal microclimate: the attorney's-fees regime, and CEQA's ability to delay projects, which gives developers an incentive to settle the cases. Litigation has the potential to be much more effective in fighting climate change than it has been; the key is for lawyers to creatively use the tools provided in their local legal microclimates.

---

[51] "China's Achievements, New Goals, and New Measures for Nationally Determined Contributions," 6–8.

[52] Rosenfeld and Sajó, *The Oxford Handbook of Comparative Constitutional Law*, 131, 172.

Legislation will be important as well. We need stronger climate laws in every country, and a strengthened global climate treaty as well, as will be discussed in the last chapter of this book.

Chapter 6
Politics

This chapter covers a range of topics concerning the climate-change political action landscape. Politics is our biggest obstacle to dealing with global heating. The next chapter will discuss what we should do, assuming that politics don't get in the way, which they will. This chapter will not discuss current politics in this area—how Republicans in the US, nationalist populists in other countries, and corporations and countries with substantial fossil-fuel interests are currently blocking effective climate action. Instead, it will discuss some of the background concepts in play in the political arguments on climate.

Terminology can be important, so I want to discuss some important terms that are used in different ways by different people, largely for political or rhetorical effect. The best example is the word "liberal." It is used by the right wing as a disparagement, referring to a version of capitalism with a large safety net, a welfare-state liberalism. But, though they don't call themselves liberals, many right-wingers espouse another form of liberalism: classical liberalism, aka neo-liberalism, which promotes individual freedom and limited government, and is pretty close to libertarianism. It's important for us to have a common understanding of what these terms mean.

Liberalism

The base meaning of "liberal" from the dictionary is "a political theory founded on the natural goodness of humans and the autonomy of the individual and favoring civil and political liberties, government by law with the consent of the governed, and protection from arbitrary authority."[1]

---

[1] "Liberalism" in *American Heritage Dictionary*. 5th ed. (Boston: Houghton-Mifflin) 2011.

"Right-liberalism" — the conservative or right-leaning version of liberalism — comes from Adam Smith and John Locke and other Enlightenment thinkers, and is focused on economic "liberty." The idea is that if individuals, and nations, will follow their self-interest, the result will be the best for the collective:

> He generally, indeed, neither intends to promote the public interest, nor knows how much he is promoting it ... by directing ... industry in such a manner as its produce may be of the greatest value, he intends only his own gain, and he is in this, as in many other cases, led by an invisible hand to promote an end which was no part of his intention. Nor is it always the worse for the society that it was no part of it. By pursuing his own interest he frequently promotes that of the society more effectually than when he really intends to promote it. I have never known much good done by those who affected to trade for the public good. It is an affectation, indeed, not very common among merchants, and very few words need be employed in dissuading them from it.[2]

Though this "invisible hand" theory is expressed in economic terms, it has bled into the political sphere by analogy, much the way the scientific theory of evolution was applied by analogy in the political sphere as social Darwinism in the nineteenth century.

Right-liberalism is essentially libertarianism, and seeks to minimize governmental interference in all aspects of human life. It promotes self-reliance and individual responsibility: each person should take care of himself or herself and be responsible for her or his safety and well-being, without any help needed from the government. Government should therefore be strictly limited to those functions that only government can provide, such as the military and the court system. Libertarians are in favor of civil rights, especially when they are formulated the way they are in the US Bill of Rights, as limitations on governmental interference, not as positive rights.

---

[2] Adam Smith, *Wealth of Nations* (London: Penguin Books, 1776).

The term "liberal," in general US political parlance, refers to social liberalism, or what I will call "left-liberalism," which gained steam in the New Deal that dealt with the great depression of 1929. Although left-liberalism favors a market economy, it recognizes that strict the laissez-faire economy advocated by libertarians is inequitable. Left-liberals support a government role in addressing economic and social issues such as poverty, health care, education, and the environment.

Progressivism

Another term with multiple meanings is "progressive," which, like "liberal," has had two significantly different meanings. The term contains the word "progress," and is thus based on the idea that there can be progress in the social sphere. The Progressive Era, during which this meaning was emphasized, lasted from the 1890s to the 1920s.[3] The movement at that time targeted government corruption by big business and sought to professionalize the government. The idea was to use scientific management and technical expertise to run government, so it would be technocratic instead of political. A thread of this type of progressivism still persists among left-liberals in the US and EU, and it's one of the ideas most hated by the right-leaning nationalists, such as the January 6 insurrectionists in the US and the Gilets Jaunes in France, who think the leftie elites look down on them, as they probably do.

This earlier meaning has now largely been forgotten, except by historians. Now, many left-leaning Americans call themselves progressives instead of liberals, acquiescing to the right-wing disparagement of the latter term.

The dialectic

The main idea behind dialectic, closely related to the word "dialog," is a back-and-forth argument, as exemplified in Plato's dialogues, put in the mouth of Socrates—hence the "Socratic method"—to find the truth. Modern German philosophers such as Fichte and Hegel extended the

---

[3] "Progressive Era" In Wikipedia, https://en.wikipedia.org/wiki/Progressive_Era

concept to focus on resolving contradictions: a thesis is contradicted by an antithesis; resolution of the contradiction results in a synthesis, which becomes a new thesis to which a new antithesis is proposed, and the process continues. This is often called the "Hegelian dialectic," though Hegel did not himself use the thesis-antithesis-synthesis terminology.

Karl Marx latched onto the dialectic in his theory of historical process, which he called "dialectical materialism." He extended the idea of the dialectic from the realm of philosophical argument into the sphere of historical reality. He theorized that history made inevitable progress toward an ideal state—the fall of bourgeois capitalism and the rule of the proletariat—through a pendulum-like back-and-forth contention of historical movements and forces. Marx turned out to be wrong: his predictions of the inevitability of a communist utopia where workers own the means of production have not come to pass, of course.

But the dialectic is still useful as a concept when analyzing history. We can see the pendulum-like back and forth in US politics, for example. Control of the White House and Congress alternates between Democrats and Republicans. When one party has been in control for a certain length of time, supporters become dissatisfied and opponents become empowered, and the other party is voted in.

Socialism and communism

Socialism was originally a movement advocated that the means of producing and distributing goods and services be collectively owned, usually by the government. In the early part of the 20th century the Soviet and Chinese Communist governments tried to run their economies on that basis, and their attempts showed that centralized government planning of the economy, where the government controlled production and distribution of goods, and planned how much of each type of goods to produce, was much less effective than a market economy in allocating resources to meet demand.

Although no nation today centrally plans its whole economy the way the Soviet and Chinese Communists did, it is common for governments to own and operate certain types of enterprises, especially ones that are seen as providing important public infrastructure. Examples include the postal system, energy production, transportation (trains and airlines), broadcasting, and health care. The US government has divested most of its commercial interests, but it still runs US Postal Service as a quasi-governmental agency. France has recently started to nationalize Electricité de France, which provides nuclear power. In 1982, when François Mitterrand's Socialist Party came into power, the national government nationalized several significant infrastructure companies, including banks, telecommunications, and electrical power.[4] Great Britain and France nationalized the railroads in the 1930s and 1940s. The US semi-nationalized passenger railroads in 1971 when it created Amtrak. The British Health Service (BHS) is socialistic – it is a government-owned and -run business providing primary healthcare to all residents. The single-payer health-insurance plan proposed periodically in the US would nationalize health insurance instead of health care and would therefore be much more limited than the BHS, but still would be socialistic in the original sense.

Socialism was not just a creature of the left. It was a strong component of several right-wing fascist programs such as Nazism. "Nazi" is short for *Nationalsozialismus*, German for National Socialism. Under Hitler, the German state nationalized or otherwise took over control of industries that were essential for the war effort.

The federal government in the US was much smaller before the 16th Amendment was added to the US Constitution in 1913; that amendment allowed Congress to impose an income tax without apportioning it among the states in a particular way. Increased taxation allows the US Government a lever of economic control via fiscal policy. Congress created the Federal Reserve in that same year, giving it a means to implement monetary policy. These economic tools were used sparingly until the Great Depression.

"Socialist" was a term of disparagement in this country in the 1920s, when business and the government lined up behind market (laissez-faire)

---

[4] "List of Nationalizations by Country," in *Wikipedia*,
https://en.wikipedia.org/wiki/List_of_nationalizations_by_country.

capitalism. Any curb on the rights of businesses to make money was viewed as un-American. During the Lochner era in the early twentieth century the US Supreme Court regularly overturned state laws regulating economic practices such as minimum-wage laws and child-labor laws.[5] They interpreted the US Constitution's substantive due-process rights as precluding any state interference with economic liberty or private contract rights. The Court did an about-face in 1937, allowing states to regulate business practices.

Right-wingers have spread the notion that socialism is an evil, like totalitarianism. In doing so, they wrongly conflate politics and economics, which can be independent of one another. The Chinese government, for example, provides limited civil rights and rigidly controls politics, ruthlessly suppressing advocacy for political democracy, while it allows a fair degree of economic freedom. You can start a small business and it can prosper in China now. In the Scandinavian countries, it's just the opposite. Civil rights and political liberties are championed, but the government is very involved in redistributing income through taxation, and the government imposes a huge regulatory burden on businesses. One could theoretically have a totalitarian dictatorship combined with laissez-faire market economics, or a democratic country with a communist economy, i.e. public ownership of businesses.

The Great Depression that started in 1929, like the smaller Great Recession of 2007, was caused in large part by the excesses of unrestrained market capitalism. Both events were part of a dialectical process, causing regulation to be enacted to protect consumers from the effects of overly risky behavior by financial institutions. Several important laws enacted in response to the Great Depression constituted the "New Deal," which was a significant departure from the laissez-faire, market economics which had been the American dogma. Such laws include the 1933 Securities Act, which regulated the sale of stocks and bonds and established the Securities and Exchange Commission, the Wagner Act to protect labor organizing, the Works Progress Administration relief program, providing safety-net employment, the Social Security Act, the Fair Labor Standards Act of 1938,

---

[5] "Lochner Era" in *Wikipedia*, https://en.wikipedia.org/wiki/Lochner_era.

which set maximum work hours and minimum wages for most workers, and laws establishing the Federal Deposit Insurance Corporation (FDIC), the Federal Housing Administration (FHA) and the Tennessee Valley Authority (TVA).

Right-wingers called these programs socialistic at the time, because they interfered with the market economy. This is the sense of the word as it is mostly used now, and it's pretty far removed from its original meaning, that the workers, or the government, representing the workers, would own the means of production.

But mostly, when libertarians and other economic conservatives complain about socialism, they are referring to government programs that help the disadvantaged, such as welfare, Medicaid, social security, and food stamps, the so-called social safety net. They brand such efforts socialistic because it is persuasive for many of their followers. Socialism = communism = totalitarianism and it takes away our freedom. Such rhetoric lacks nuance, partly because it ignores the distinction between the economic and the political.

Zero-sum game or labor theory of value?

Classical economists and David Ricardo posited that the value of a thing comes from the labor required to produce it.[6] It is ironic that right-liberals espouse this theory, which played an important part in the economics of Karl Marx. Right-liberals argue that wealth should belong to those who created the wealth through their labor. Bill Gates' share of Microsoft stock, for example, represents the value he provided by creating Microsoft, which provided tremendous value to the world. Since Gates created this wealth himself, he should be allowed to keep all of it and control how it is used. I would offer two criticisms of this argument.

First, most of the things we build are dependent on social, economic, and physical infrastructure provided by our society at large. Microsoft depends

---

[6] Graham Bannock and R.E. Baxter, "Theories of Value," in *The Penguin Dictionary of Economics* (London: Penguin Books, 2011), 399.

on roads and public transportation to bring its employees to work, and to ship the physical products they make. Microsoft depends on the legal system to enforce its contracts. Microsoft depends on the banking system for credit and other financial services. And so on. No company is an island. Any enterprise that creates wealth is dependent on the complex infrastructures within which it is embedded.

Second, a significant part of our financial and political system functions as a zero-sum game, where one player's gain must be equal to other players' losses. Some people, like our former President Donald Trump, are good at grabbing more than their fair share of life's economic goodies. They get what they have by taking from others. Is it fair that the reward for winning this type of game should be so large? I would say it isn't, and that we should use our tax system to redistribute some of this wealth to others less talented at grabbing more than their fair share, because that would be more equitable.

Environmental justice

There's a sense in current environmental circles that racial issues, social justice, and identity politics trump everything else, including the environment. When these issues come into play they foreclose all other considerations, including trade-offs between them and the environment. In many circles, it is not possible, for example, to have a reasoned discussion on whether the US should curtail immigration into the country on environmental grounds. The idea is labelled "racist," and that's the end of it. But if, as discussed in Chapter 3, population is a major factor in our environmental problems, the US should want to set an example to the rest of the world by stabilizing its population. The US population would be stable, or perhaps slightly in decline if it were not for immigration.[7] We can't have a rational discussion on how to stabilize our population without taking immigration into account, but doing so is deemed racist by many environmentalists.

---

[7] "Census Finds U.S. Population Will Decline Without Immigration," Cato Institute, Feb. 14, 2020, https://www.cato.org/blog/census-finds-us-population-will-decline-without-immigration.

There's also a strong attachment to environmental justice in the environmental community now. Many environmentalists believe that it is impossible to separate social-justice concerns from environmental concerns, so we can't properly deal with climate change without changing our society to eliminate racism, colonialism, white supremacy, etc. I've heard this argument many times, but have never been convinced. If we can stop burning fossil fuels and stop deforestation, we'll have done most of what we need to do on climate change. There are significant distributional and justice issues when doing these things. It's important that the burdens not fall on those least able to deal with them. It's important that we not exacerbate social inequities. But we don't need to eliminate all social injustices to deal effectively with the climate.

Are corporations more powerful than States?

This chapter focuses on governments of States. But a legitimate question is: how important are States? The last century has seen the rise of large corporations that have more resources than the governments of many countries.

The US has 2.1 million civilians working for its national government, about the same number as the larger corporations have employees: Amazon (about 2.2 million), McDonald's (1.9 million), and Walmart (1.7 million). Only nine countries have government revenues larger than Walmart. So the largest corporations have a scale of operations roughly the same size as governments. Corporations have much more freedom of action: they are responsible to their stockholders and directors, but not to a political electorate. Unfortunately, the law in most jurisdictions requires corporations to put the financial interests of their shareholders above all other interests; they are not allowed to reduce their profits to help fight climate change if it is not in their economic best interest.

Many corporations are touting their plans to achieve net-zero GHG emissions within a decade or two. In many cases this amounts to greenwashing; American environmentalist Bill McKibben joked that net-zero for Exxon would mean they drive all-electric Ford F150 trucks, instead of gas-powered Ford F150 trucks, around their oil-production and refining facilities. There is a wide disparity in what corporations need to do to respond to the climate emergency. Low-impact businesses like law firms can maintain offices in LEED-certified buildings powered by renewable energy, minimize flying, encourage the use of transit for commuting, and offset their remaining emissions. High-impact businesses like the oil majors need to go out of business, or at least to get out of their current business. Businesses will not lead on climate because a proper response to global heating is an existential threat to the businesses that most need to change.

The largest corporations, like Apple and Google, are international. They can take advantage of loopholes in national laws by locating their operations where governments will impact them the least, e.g. in countries with low tax rates or the laxest environmental laws. They have money to lobby governments for favorable laws and regulations. It seems to me that, over the course of my 70-year lifetime, the power of the largest corporations, compared to governments, has substantially increased.

Do we love where we live?

My wife, and several of my friends, have talked about moving to a country other than the US as the result of recent political developments in this country. My wife would do it as a protest, to show she is not complicit in the policies our government is pursuing. But moving for this reason wouldn't be worthwhile—who would notice or care?

We live in California, the country's most populous state, and a state that is in many ways like a country. The state has a GDP roughly equivalent to that of France, though our population is two-thirds of France's. Our environmental laws and policies are much better than those of other states; they're roughly equivalent to those in the EU. The state leans left-liberal. One frustration in living here is that our votes count less than other US residents, due to quirks in the US political system. The electoral-college system makes presidential elections winner-take-all in most states. Since California will almost surely support the democratic candidate in a general presidential election, our votes for president don't count as much as voters in swing states. Eliminating the electoral college in favor of a popular-vote election for president would eliminate this anti-democratic situation.

My wife and I could move to another state where our votes would count more. In the US that would be a swing state. Moving to a deep-red state would nullify our votes just as much as they're nullified in a deep-blue state like California. If we moved outside the US, it might take a number of years to get the right to vote in the new country. But where would our vote do the most good? China is the most important country for dealing with climate change, because its emissions are the largest of any country and are still increasing. But votes in China are not valuable for causing a change in course on climate because the Chinese government's policies aren't significantly influenced by election results. Maybe India, the third-largest emitter, would be the place to go to get a vote that counts the most for the climate.

Most of us don't choose where to live based on such a narrow single criterion. We choose based on a wide range of factors, such as the availability of appropriate employment, the weather, cultural amenities, and the cost of living. We also tend to value the company of like-minded folks. I'm pleased to be part of a strong legal and activist community fighting climate change in California. This community is bigger, stronger, and better integrated into mainstream state politics than an analogous community would be in Mississippi, for example.

Such communities don't need to be place-based. Covid has shown us that we can easily interact with people far away, and can replace many in-person meetings with calls on Zoom or other interactive Internet platforms. By choosing which online communities to join, we can select the people we want to be involved with. This is similar to the way we can "friend" or "follow" others on social media, thereby focusing on the like-minded.

Could we also select the laws we want to live under this way? Ever since I learned, in law school, that the states in the US could have wildly different legal systems I've wondered why they don't. Why isn't there a state that's very left-liberal, which taxes and redistributes income at a high rate, like a Scandinavian country? Why isn't there a libertarian state, with very low taxes, small government, and limited regulation?

In a recent book, *The Network State: How to Start a New Country*, American entrepreneur Balaji Srinivasan suggests that a properly motivated group could build a State online by organizing the group online, fund-raising to acquire parcels of land which would eventually become the country's territory, setting up a cryptocurrency and system of smart contracts as the basis for an economy, and eventually seeking diplomatic recognition from established States.[8] As far as I can see, he pretty much ignores the issue of how such a state's legal and legislative systems would work, relying on the general notion of "smart contracts" to manage everything. He wants to replace big government with decentralized network applications built on blockchain technology.

Such a network state is highly unrealistic and impractical, but the concept provokes the question of how we could universalize some aspects of our lives that are currently determined by the geographical point on Earth where we live. The accident of our birth in a particular place determines a great deal about our life, more than is fair, even though most people end up fairly happy with where they live. A 2018 Gallup poll found that only 15% of adults would migrate to another country, if given the chance to do so.[9] We're mostly comfortable with the people who live around us because they tend to be culturally similar. And a network state would be impractical for the one-third of humanity that has never used the Internet.[10]

How religion is taken into account in the way a country is run one of the most important factors distinguishing countries' governments. My view is that religion should be kept out of government. Individuals deciding what should be done will of course be guided by moral considerations and, for many, that morality comes from religion. But I like the French approach, called laïcité, under which religious concerns are relegated to the private

---

[8] Balaji Srinivasan, *The Network State: How to Start a New Country*, 2022, https://thenetworkstate.com.

[9] Neli Esipova, Anita Pugliese, and Julie Ray, "More than 750 Million Worldwide Would Migrate If They Could" (Gallup, December 10, 2018), https://news.gallup.com/poll/245255/750-million-worldwide-migrate.aspx.

[10] International Telecommunications Union, "Facts and Figures 2021: 2.9 Billion People Still Offline" (ITU, November 29, 2021), https://www.itu.int/hub/2021/11/facts-and-figures-2021-2-9-billion-people-still-offline/.

sphere, and must not serve as the basis for government policies. This goes further than the US Constitution, which prohibits a state-established religion, but leaves enough ambiguity that many in the US contend that the US is a Christian country. The far end of the spectrum is occupied by religious states like Pakistan and Israel that incorporate religion into government. I'd never want to live in such a country.

Some things ought to be universal, like human rights. We should strengthen international law so that it covers more of these universal areas, and provides rights that may be enforced by individuals. I don't think we'll ever have a non-place-based country because too many of the services government provides, like police, building regulation, and infrastructure, are place-based. But we can and should move in the direction of increasing and harmonizing international law relating to human rights and climate change, and perhaps in other areas.

Internationalization v. nationalism

I used to work for a law firm whose biggest client was the National Rifle Association (NRA), so I came into contact with gun nuts. They love their guns and obsess about them. One trope I encountered there was the invasion of the US by international forces such as the United Nations, with black-uniformed UN soldiers parachuting from black UN helicopters into small towns in the heartland. The 1984 movie *Red Dawn* showed a group of teenaged guerillas resisting such an invasion by the Soviet Union and its allies. Firearms enthusiasts use these scenarios to rationalize their need to own highly effective weapons like AR-15s. They see themselves as akin to the original American revolutionaries who used guerilla tactics to defeat and drive out the British. They see international entities as the enemies of US sovereignty and values.

Some extreme Christians believe the end times described in the Book of Revelation are at hand. My father, during his childhood in a strict Baptist family, spent some of his Sundays swinging in the swing in the back yard waiting for Jesus to appear in the clouds for the second coming. (People with this viewpoint don't tend to worry about climate change because they think the world will end before it can get too bad, or at least that God must be watching out to make sure it doesn't get too bad.) Revelation suggests that the Antichrist, working at the behest of Satan, will preside over a global government during the end times, so extreme Christians think we should be wary of an international government because it could be the instrument of the devil.

The idea of the "international" as a bogeyman threatening national culture and interests goes back at least a few hundred years. Jews are the prototypical international bogeymen the National Socialists (Nazis) used as the opposite of their "blood and soil" nationalism. They thought the Jews, spread out in all countries, ran a spidery international financial network that controlled the world. These thoughts echo strongly in current antisemitism, such as that of the extreme right in the US and Europe.

It is legitimate for citizens to want to preserve their national cultural characteristics. Ideally, state boundaries would be based on them. France is a territory where most people speak French and identify with French cultural traditions going back hundreds of years. Drawing boundaries based on such cultural characteristics is an imperfect science, though, and subject to political considerations. For example, about 25 to 30 million Kurds live in communities in Iran, Iraq and Turkey.[11] This number is similar to the population of Australia, and deserving of a country, but the countries where the Kurds live do not want to cede territory for a Kurdish nation. Another example: are there enough Catalan people in Spain to justify creating a new nation for them? The Catalan seem to think so, though Spaniards as a whole disagree.

But nationalists around the world take this idea too far, and bristle at any limitations on their sovereignty. They want to be able to do whatever they want inside their country, even if it violates international laws and norms. Too often, though, the "whatever they want" is not democratically determined and doesn't represent the will of the country's people as a whole, but rather the will of a self-interested minority that has the support

---

[11] "Kurd People," in *Encyclopedia Britannica*, November 28, 2022, https://www.britannica.com/topic/Kurd.

of those in control. Appealing to nationalism is often an appeal to us-versus-them tribalism, which can be very effective political rhetoric.

The US has been especially resistant to agreeing to participate in international efforts, particularly those relating to human rights. There's a strong strain of isolationism going way back in the US. According to the American Civil Liberties Union (ACLU), the US has ratified or acceded to fewer key human rights treaties than any other G20 (top 20 world economy) country.[12] The US is the only country other than Somalia to fail to ratify the Convention on the Rights of the Child, and is one of only seven countries that has failed to ratify the Convention on the Elimination of All Forms of Discrimination against Women.[13] It has withdrawn from the Rome Statute of the International Criminal Court, which established an international criminal court that prosecutes individuals for genocide and war crimes. On the environmental front, the US is one of just two countries that have not ratified the Convention on Biological Diversity, a treaty that protects endangered plants and animals around the world. It hasn't ratified the Stockholm Convention on Persistent Organic Pollutants, or the UN Convention on the Law of the Sea. The Council on Foreign Relations attributes the US' reluctance to participate in international law regimes that have been accepted by most other countries to "a stalwart commitment to a narrow conception of national sovereignty and to the ideal of American exceptionalism."[14]

In declining to participate broadly, let alone lead, in international agreements on important issues, the US is acting against its own best interests, and the interests of the world at large. Hawks and isolationists tend to view international relations through the lens of competition with other countries, but it's better for everyone to emphasize cooperation over

---

[12] "Treaty Ratification" (American Civil Liberties Union, 2022), https://www.aclu.org/issues/human-rights/treaty-ratification.

[13] "United States Ratification of International Human Rights Treaties" (Human Rights Watch, July 24, 2009), https://www.hrw.org/news/2009/07/24/united-states-ratification-international-human-rights-treaties.

[14] Anya Wahal, "On International Treaties, the United States Refuses to Play Ball" (Council on Foreign Relations, January 7, 2022), https://www.cfr.org/blog/international-treaties-united-states-refuses-play-ball.

competition. As globalization increases, there are more and more global problems that need international solutions. A good example is the UN Convention on the Law of the Sea (UNCLOS), which has been ratified by 168 countries. It provides a legal framework for marine and maritime activities. The US has neither signed nor ratified this treaty, because of concerns that it would undermine US sovereignty by transferring "ownership" of the high seas to the UN.[15] The US' failure to participate in such efforts will, if it continues, be very harmful in the climate sphere, where new treaties requiring international cooperation will be required, as discussed in the next chapter.

In a broader sense, moving toward an international legal system would be advantageous. I estimate that about half of our current laws could be universal, the same in every country, without causing friction with local customs and usages. Moving in that direction would be generally beneficial; moving in that direction is necessary to deal with the climate emergency, as discussed in the next chapter. Human rights law should be the same everywhere. And why should the law of contracts vary from place to place? A universal, international law of contracts could benefit international commerce, as trade becomes ever more global. For such laws to become universal, they need to be agreed to in treaties ratified by all nations, and all nations must agree that they are enforceable as local law within all nations. It will benefit everyone to move toward non place-based laws and systems—it will provide best practices for everyone, and help insure that everyone has the same basic rights.

The decline of the US

The idea has been around for a long time that civilizations rise, flourish, and decline, like the birth and death of people. The Roman Empire is often used as an example: Rome, a plucky little city, was bursting with young energy when it conquered Italy, then North Africa, Gaul (France and Germany) and the British Isles. It got old, rich, complacent, and decadent, and couldn't stand up to the invasions of the barbarians at the gates, so it was conquered in turn and the empire ceased to exist.

---

[15] Wahal.

Arnold Toynbee, an academic historian, and one of the most famous public intellectuals of the early twentieth century, developed a theory of history under which civilizations arise in response to a challenge, flourish for a while, and then die, usually from suicide. A series of nations in modern Europe have held the ascendancy in turn: the Netherlands, Portugal, Spain, France, and then England. The two World Wars were about whether Germany would be next to be dominant, and the answer was no—the next dominant power would be the United States of America.

The US is dominant now. It's the top economy, measured by GDP, though China is ahead if the economy is measured by purchasing power parity. And US military expenditures are about twice the combined expenditures of China, Russia, Saudi Arabia, and France, the four countries with the next-largest militaries.

The US came into its own as a world power as a result of World War II. It won the war by scaling up a high-quality military and supporting it with a huge production of materiel. After victory in, essentially, two wars fought in parallel—the war in Europe then the war in the Pacific—Americans were empowered to use the abilities that won the war to greatly expand the peacetime economy. That moment, right after World War II, was an apogee for the country. Americans were excited by what they'd accomplished, energized, willing to sacrifice and work to build.

Henry Luce, an important magazine publisher, declared the start of the American Century following World War II. America's performance in that war showed we could beat any adversary in war through a combination of will and ability to fight and industrial capacity. We were also indisputably the good guys in that conflict, given the imperial ambitions and horrors inflicted on conquered peoples by Germany and Japan

But now there are only twenty-some years left in Luce's American Century. The US does have the largest economy of any country, as well as the most powerful military. But we may have passed the peak. Analysts are divided about when China's economy will grow to be larger than that of the US. Our military dominance is fueled by our large economy; it won't last forever. There's no reason to think that the usual pattern we've seen over thousands of years should not continue. The US will lose its dominant position to the next rising dominant country, which will probably be China.

American exceptionalism

There is a view that the US is special, a beacon of liberty and democracy illuminating the way for other countries, leading the implementation of Providence, the Christian God's plan for humans on Earth. Evidence for this is on the US one-dollar bill, which has a very strange circular seal with an eye radiating light in a little triangle on top of a pyramid. The motto "annuit coeptis" is above the pyramid, and "novus ordo saeclorum" is below it. The first of these phrases translates as "He [God] has favored our undertakings," meaning that the US is providential, carrying out God's plan. The second phrase means "the new order of the ages," i.e. American exceptionalism. Above the word "ONE" on the bill are the words "In God We Trust." We will ignore this Christian view of American providence when deciding if the US is a special country, unlike others.

It may be true that the US was the first modern democracy with a written constitution, adopted in 1789. The US Constitution preceded the first constitution adopted under the French Revolution by two years. Some Americans laud the wisdom of the Founding Fathers who drafted the US Constitution in 1787–1788, but those men added only a bit of intellectual content to what came from Enlightenment thinkers in France. In intellectual terms, the French Revolution and its Enlightenment background in Voltaire, Rousseau, Diderot, etc., is much richer.

The EU is the system most comparable in legal structure to the US. Both are democracies with a federal system, and roughly the same size. The EU is the previous dominant power, at least if we include the UK and its colonial empire, which came apart during the time of the two world wars. The EU is arguably more democratic, given the problems with the US Constitution discussed in the previous chapter. The US has more permissive free-speech rights, a good thing in my opinion. Americans make more money than Europeans — the US per-capita GDP is 60% higher than the GDP in France. But a big factor causing this difference is the European choice to work less and consume less, so as to have more time for leisure. And religion has less sway in political decision-making in Europe. When I visit EU countries, life there seems very similar to life in the US.

So I can't see that there is any basis in fact for the view that the US is an exceptional country, beyond its currently being the dominant country, economically and militarily. Much of the American exceptionalism is fueled by tribalism: "Rah my team!," and by ignorance: over 50% of Americans have never owned a passport; they don't realize how similar life in other countries is to life in the US.

Who are the bad actors?

Who's holding up action on the climate? Countries and corporations that make most of their income from selling fossil fuels, like Russia and Exxon, want to stave off climate action so they can continue making money as long as possible.

I've yet to see a convincing analysis of the geopolitical changes that will result when we stop burning fossil fuels. The Middle East will be of much less strategic importance once we stop buying oil produced there. Russia will sink further into unimportance.

Saudi Arabia: It's a repressive country that's also the world's largest oil company. At the 2022 United Nations Climate Change Conference (COP27), Saudi Arabia pushed to block a call for the world to burn less oil, saying that the summit's final statement "should not mention fossil fuels."[16] Its government-controlled oil company, Saudi Aramco, produces about a tenth of the world's supply of crude oil. Saudi Aramco has funded over 500 studies aimed at keeping gasoline cars competitive or casting doubt on electric vehicles. They also pushed to delete from an IPCC report a sentence calling for an active phaseout of fossil fuels.

---

[16] Hiroko Tabuchi, "Inside the Saudi Strategy to Keep the World Hooked on Oil," *New York Times*, November 21, 2022,
https://www.nytimes.com/2022/11/21/climate/saudi-arabia-aramco-oil-solar-climate.html.

Oil Majors: They, and Exxon in particular, have spent large on climate disinformation campaigns to sow doubt about climate science: Is climate change real? Is it caused by human activities? In the distant past, Earth's climate varied widely, and we're just seeing the same natural processes play out. The campaign is based on the very successful disinformation campaign by tobacco companies that cast doubt on science showing the health impacts of smoking. They're doing this to keep making huge profits from selling gas and oil, and want to keep doing that for as long as they can.

Motivation for change

The Japanese attack on Pearl Harbor on December 7, 1941, motivated the US to enter World War II. That attack threatened the US' geopolitical interests in the Pacific, but also invoked the US' tribalism. We were suddenly attacked by a foreign tribe, with no warning, provoking a huge emotional response. Climate change has been likened to the mythical frog that is placed in a pan of cool water that is then brought to a boil. Supposedly the frog never leaps out because the heat comes on gradually. We're seeing climate impacts increase in a similar gradual way. Each year we break temperature records, have more days hotter than humans can tolerate, suffer more intense rainstorms, hurricanes, and wildfires than in the past, and watch Arctic and Antarctic sea ice shrink. It happens gradually, and we're not always aware of the contribution that global heating makes to these impacts. This graduality motivates us much less than a sudden attack like Pearl Harbor.

Some very strong political and economic actors are opposed to taking action on the climate. The oil majors, including several Middle-Eastern countries and Russia, and other business interests want to keep making huge profits while they still can, so they sow disinformation, and engage in extensive political lobbying to prevent strong action in this area.

The fact that the climate emergency is caused by many peoples' actions and affects everyone in the world also saps our motivation. We would be more motivated, perversely, if it was caused by another tribe and affected just us or our tribe. It impacts non-human animals and future generations, groups we're not accustomed to including in our moral calculus. All of these demotivating factors coming together in this very important issue result in what one writer characterizes as "a perfect moral storm."[17]

---

[17] Stephen M. Gardiner, *A Perfect Moral Storm: The Ethical Tragedy of Climate Change* (Oxford: Oxford University Press, 2011).

Chapter 7
What We Need to Do

I hope this book has convinced you that climate change is the most important emergency facing the world now. My opinion is that it is the most important challenge humans have ever faced. It's more important than the problems we faced in World War II. It's more important than human rights, racism, or economic inequality. We need to be on a war footing. Climate is more important than terrorism and drugs, which we tried to deal with, in the US at least, in the "war on terrorism" and the "war on drugs." We need a war on global heating.

What we're doing now on global warming feels like a path to failure. We keep paying lip service to the 1.5°C goal for political reasons, even though we're past the point where we could achieve it. The 2°C goal is slipping away as we dither around. In this chapter I describe the path we need to take, without regard for the politics of getting global agreement on the path. There are many political obstacles to "solving" global heating.

But it's valuable for supporters of taking strong action in this time of crisis to consider what is required to deal with the crisis, in order to compare what we're actually doing with what we need to do. This comparison will highlight the shortfalls and show where we need to take stronger action.

Individuals can take action to reduce their climate footprints, such as by trading in their gas-powered vehicle for an electric car, by switching off natural gas at home, by installing solar panels and batteries in their homes, by conserving energy and water, and by eating less meat. But these actions, even if they were undertaken by everyone in the world, fall far short of what's needed to deal with the crisis. Strong action by governments and corporations is essential.

If we could negotiate a strong international treaty dealing with climate change, States would give up some of their sovereignty and right to independently decide about climate policies in order to participate in a global plan to cooperatively deal with the climate emergency. If such a treaty were broadly adopted, it would provide political cover for adopting politically unappealing (but necessary) measures domestically within countries, and could provide a buffer against states changing their climate policies and commitments because of domestic politics. The treaty proposed in this chapter is modelled on the TEU and TFEU treaties that established the modern EU: countries gave up some sovereignty in order to establish cooperation in a number of areas.

Ideally, we will not just get to carbon neutrality, we'll restore the climate to what it was in pre-industrial times by reducing GHG concentrations to around 350 p.p.m. Nobody's talking about this now because the task of achieving carbon neutrality is itself quite daunting, and we need to focus on that.

We need to move fast. I have little doubt that we will get to near-net zero emissions eventually, because climate impacts will keep getting worse until we get to net zero. But we need to get there quickly, to minimize the impacts on future generations. Canadian environmental scientist Vaclav Smil points out that global energy transitions, such as from wood to coal, and coal to oil, have taken 50 to 60 years in the past; and he doesn't expect the transition to renewable energy to go any faster.[1] But those transitions were driven entirely by private economics; homes and enterprises switched energy sources when it made financial sense for them to do so. By contrast, we have climate change forcing our hand, and a strong societal need to transition quickly. We can speed up the transition by agreeing to adopt enforceable policies to speed up the adoption of renewable energy.

We should consider ourselves to be on a war footing. If we can declare wars against drugs and terrorism, we can declare a war against global warming. It's the most important problem facing the human race today, and probably the most important challenge humans have ever faced.

---

[1] Vaclav Smil, "A Global Transition to Renewable Energy Will Take Many Decades," *Scientific American,* January 2014,
https://www.scientificamerican.com/article/a-global-transition-to-renewable-energy-will-take-many-decades/.

Axioms

In mathematics, axioms are the assumptions that form the starting point for the elaboration of a mathematical system. Euclid, the ancient Greek mathematician who lived around 300 BCE, built his system of geometry on five axioms, one of which states "that all right angles are equal to one another." Truths deduced from the axioms are called theorems. In this section I'll set out the axioms we'll use to figure out what we need to do to optimally respond to climate change. The previous chapters of this book provide support for these axioms.

The premise of this chapter is that the people of Earth have realized that climate change is the number-one problem facing this planet, and that action to stop and, ultimately, reverse its impacts is required. We're on a war-time, all-hands-in-deck footing, and we all agree that we must make individual and collective sacrifices to deal with this catastrophe. This is not an axiom but a scenario that helps to set the scene. For people to be willing to accept sacrifices in the name of climate, there will need to be a much higher level of urgency than we are seeing now.

Axiom 1: We will take whatever action is required to keep global average global temperature increases below 2°C. This includes reaching net zero GHG emissions by 2050 and a substantial reduction in emissions by 2035. It would be nice to eventually reduce GHG concentrations to pre-industrial levels below 350 p.p.m., but we can wait to do that.

Axiom 2: We will not try to solve every other problem in the world on the back of fixing climate change. Dealing with the climate emergency must take precedence. We won't eliminate global poverty or hunger; we won't eliminate capitalism; we won't eliminate racism, colonialism, and white supremacy; we won't eliminate environmental injustice.

Axiom 3: We can't make any of the problems listed in the previous axiom worse with our actions. Distribution of the impacts and costs must be fair. There will need to be a bargain between the rich and the poor: though the poor will probably be harmed more by global heating, rich people will have to pay more money to solve the problem.

Fossil Fuels

The most important component of our program will be to stop burning fossil fuels. There may be small exceptions for uses where there are no substitutes, but they must be balanced by GHG removal in an equivalent amount. This will effectively put the oil, gas, and coal industries out of business. Their reserves and fossil-fuel infrastructures, in which they've invested trillions of dollars, will become valueless.

To be fair and equitable, we must provide just transitions for those employed in fossil-fuel industries. Displaced employees must have access to good training for new, well-paying jobs, ideally in the renewable-energy sector. Countries, provinces, and cities that are dependent on fossil-fuel businesses must be helped to transition their economies. The fossil-fuel companies themselves deserve no help with the transition. They have known about global heating for decades and have spent many millions of dollars on disinformation campaigns to delay climate action, causing great harm to everyone.

Outlawing the burning of fossil fuels will upend the international order. Countries whose economies are heavily dependent on revenues from oil and gas, such as Saudi Arabia and Russia, will need to find other industries on which to base their economy. These countries have also known about climate change for decades and have acted to postpone any action so they can continue selling oil and gas as long as possible, so they also deserve no help with the transition.

When we end the combustion of fossil fuels, we won't completely stop producing them because they're used in ways that do not involve combustion, such as manufacturing plastics, polishes, and as waxes, solvents, and lubricants. And natural gas is an important ingredient in manufactured fertilizer. But oil and gas production will be a small fraction of what it is today.

Forestry and agriculture

We must stop deforestation world-wide, and start managing our forests to sequester as much carbon as possible. We must also adopt climate-friendly agricultural practices. There is an abundance of research on soil carbon sequestration and climate-friendly agriculture, but we still haven't established the best practices in these two areas; more research is urgently required.

Research and development

We need to solve a lot of big problems for which we currently have no solutions. There is no accepted methodology to manage forests or agriculture to optimally sequester carbon, and it needs to be developed. We need a way to produce concrete and steel with zero, or at least drastically reduced, GHG emissions. We need to develop a way to power airplanes without burning non-renewable fossil fuels. We need to try to come up with an economical way to remove CO_2 from the air. These problems all require basic research.

We also need development in areas where we've developed technology that might help the climate, but aren't yet ready for widespread use or commercialization. Examples are carbon capture and storage (CCS) for powerplants and industrial facilities, battery technology, electrification of large trucks, and city planning resources. "Technologies exist to take all industry sectors to very low or zero emissions, but require 5–15 years of intensive innovation, commercialization, and policy to ensure uptake."[2]

Help for developing countries

The countries with the highest cumulative GHG emissions are, in order by total emissions, the United States, China, Russia, Brazil, Indonesia, Germany, India, United Kingdom, Japan, and Canada.[3] These countries are

---

[2] IPCC WGIII, "Climate Change 2022: Mitigation of Climate Change" (IPCC, 2022), 106, https://www.ipcc.ch/report/ar6/wg3/.

[3] Simon Evans, "Which Countries Are Historically Responsible for Climate Change?" (Carbon Brief, May 10, 2021),

responsible for the vast majority of emissions so far. Of these, China, Russia, Brazil, Indonesia, and India can be regarded as developing countries; the others are developed countries. The lion's share of emissions in Brazil and Indonesia come from deforestation, while GHG emissions in other countries are mostly from burning fossil fuels.

On the other end of the spectrum are low-income developing countries that have not been heavily industrialized. The World Bank lists per-capita GHG emissions from Somalia (which had a 2021 per-capita GDP of USD \$446) as 0.0. The World Bank list[4] shows 55 countries, all very poor, with annual per-capita GHG emissions under 1 MTCO$_2$e. People in these countries are affected by global heating just as much as those in rich countries, but they have far fewer resources to deal with the resulting problems.

We need to provide help to the poorest countries. They want to, and will, develop their economies, and it will be to our advantage to help them do so in a way that minimizes their global-warming impacts. They may need assistance establishing environmental and climate laws, and training their judiciary about environmental issues. They may need technical assistance to develop renewable energy sources instead of fossil-based ones. And they may need loans and subsidies to finance clean development.

They should also be given reparations ("loss and damage") for the climate impacts they've suffered and will continue suffering for a long time, which were caused by the large GHG emitters.

Global action

Climate change is an instance of the "tragedy of the commons." In a famous article,[5] the population ecologist Garrett Hardin posed a hypothetical: A number of cattle herdsmen are allowed to use a common pasture to feed their

https://www.carbonbrief.org/analysis-which-countries-are-historically-responsible-for-climate-change/.

[4] World Bank, "CO$_2$ Emissions" (World Bank, 2020), https://data.worldbank.org/indicator/EN.ATM.CO2E.PC

[5] Garrett Hardin, "The Tragedy of the Commons," *Science* 162, no. 3859 (December 13, 1968): 1243–48, https://doi.org/10.1126/science.162.3859.1243.

cattle. Each herdsman, deciding whether to buy another cow to keep on the common, looks only to his own interest. Adding a cow gives him additional income; the cost of the additional impact on the common pasture from an increase in overgrazing is born by the community at large, so the herdsman's individual share of the damage is small. The result is that all the herdsmen keep adding cattle to the pasture until overgrazing reaches a critical point where the policy of allowing free access to the pasture must be revised.

Earth's atmosphere and climate system are global commons, shared by all. Individuals, corporations, and countries emit GHGs as a byproduct of activities that benefit them; these activities harm the atmospheric global commons. The harm to the emitters individually as a direct result of their activities is small, but collectively they do a great deal of harm to everyone.

In Yale economist William Nordhaus' book, *The Climate Casino* a graph shows an estimate of global costs of mitigation for two scenarios: (1) where all countries participate fully, and (2) where only countries responsible for half of current GHG emissions participate.[6] The graph estimates that it will cost 1.5% of GDP to limit warming to 2°C under the first scenario, but it will be impossible to limit warming to 2° with just 50% participation. Under the second scenario, the best that can be achieved is limiting warming to 3.5°C, at a cost of 2% of GDP. It is therefore essential that the largest emitting countries, China, the U.S., and India, participate in a program to reduce GHG emissions to zero by 2050, and other emitting countries must participate as well.

The current Paris-agreement scheme, under which countries independently decide and submit Nationally Determined Contributions (NDCs) is not working. Countries are not ambitious enough in their NDCs, and are failing to properly implement them. The NDCs that have been submitted have not put us on a path to net-zero emissions by 2050, or to a 45% emissions reduction by 2030.[7]

---

[6] William Nordhaus, *The Climate Casino: Risk, Uncertainty, and Economics for a Warming World* (New Haven: Yale University Press, 2013), 177.

[7] Peiran R. Liu and Adrian E. Rafferty, "Country-Based Rate of Emissions Reductions Should Increase by 80% beyond Nationally Determined Contributions

We can't get there without a new agreement among at least the largest emitters to do more.

Global climate treaty

I propose a new multilateral treaty, the Global Climate Treaty (the GCT), to complement the UNFCCC and Paris Agreement. It would set up an organization in the United Nations, under the United Nations Environmental Program (UNEP) which I will call Climate Control. This organization would be merged with other UN climate programs.

Parties to the GCT, which would ideally be all the states in the world, would agree to reduce their GHG emissions to net zero by 2050, and agree to phase out burning fossil fuels and deforestation by 2050. Trajectories— the reduction paths for phasing out fossil fuel and deforestation, and other actions that must be taken to limit climate damages—would be established for each country by Climate Control. The GCT parties would agree to abide by regulations established by Climate Control, and to enforce the GCT and Climate Control regulations as domestic law.

Parties to the GCT would also agree not to engage in climate-related geoengineering (using techniques such as dispersing sulfur dioxide into the atmosphere to reflect more sunlight back into space), or allow their residents to do it, without approval from Climate Control.

The GCT would establish an International Environmental Court, which would hear disputes under the GCT, and would have the power to enforce the GCT and its regulations. It should include citizen enforcement, so that NGOs could bring enforcement actions against States that fail to comply.

Fossil fuels under the GCT

In the GCT, state parties will agree to phase out burning fossil fuels by 2050. A small number of exceptions to this rule may be provided by Climate

---

to Meet the 2°C Target," *Nature Communications Earth & Environment* 2, no. 29 (February 9, 2021), https://doi.org/10.1038/s43247-021-00097-8.

Control for uses with no viable alternatives. State parties will also agree to phase out production of coal, oil, and gas by 2050, except for non-combustive uses. This production will also be regulated by Climate Control. Climate Control will also monitor each country's trajectory prior to 2050 to make sure it is on the required path. It is important to make good progress on emissions reductions before 2050; we can't just wait until 2049 and then ban fossil fuels. The quick transition to net zero would wreck the global economy.

The GCT should include provisions for the establishment of a global fossil carbon market. Such a cap-and-trade scheme would be better than a command-and-control regime for restricting fossil carbon emissions because it would allow economic markets to decide on the allowed residual uses of fossil fuels. Climate Control would decide each year the maximum allowable amount of fossil-fuel emissions for each country, and would sell allowances in that amount at a price that matched the current social cost of carbon. Available allowances would decrease each year until the net-zero year of 2050, when the number allowances would be limited to the amount of GHGs removed from the atmosphere by direct air capture or other carbon removal processes.

The State parties would be required, under the treaty, to pass the cost of the allowances to the persons burning fossil fuels. For drivers of fossil-powered cars, the allowance cost would be added to the price of gasoline or diesel fuel, for example.

The sale of allowances would provide funding for Climate Control. If Climate Control set the allowance price at the $51/ton social cost of carbon recently adopted by the U.S. government, the sale of allowances for the 59 GtCO₂e currently emitted each year would bring in USD $3 trillion. Climate Control's fossil carbon market would preempt other cap-and-trade systems, such as the ones run by California and the EU.

Such a cap-and-trade system is preferable to a global carbon tax because it will be able to absolutely limit emissions as we approach the year 2050. It will limit emissions by limiting the sale of allowances. The closest a carbon tax could come to this would be to tax emissions at a very high rate, but there's no way to exactly predict how much emissions reductions will result from a given tax rate.

A treaty that has already been proposed, the "Fossil Fuel Non-Proliferation Treaty" would require state parties to phase out their production of fossil fuels.[8] This, unlike most mitigation proposals, attacks the problem from the production, rather than the demand side. It is important to regulate production as well as consumption. As documented in the United Nations' Production Gap Report,[9] governments still plan to produce much more coal, oil, and gas than we can afford to burn, if we are to limit global temperature rise to 2°C. This signals that governments are not serious about reducing fossil-fuel emissions.

The Fossil Fuel Non-Proliferation Treaty may turn out to be the first step toward the GCT. The GCT needs to be much broader than the Fossil Fuel Non-Proliferation Treaty, and needs to have enforcement provisions that will make it binding and effective in ramping down fossil emissions to net zero.

Deforestation under the GCT

The GCT would require an end to net deforestation much sooner than 2050. Ramping down fossil-fuel use will require a massive reorganization of the energy sector, and can't be done quickly. Deforestation, on the other hand, is driven mostly by clearing forest land for agriculture or other purposes, and could be stopped quickly without huge adverse consequences for the economies of Brazil and Indonesia, the countries where GHG impacts of deforestation are greatest.

Command and control would work much better than market-based mechanisms in this area. The GCT should prohibit net deforestation in each state party from the date of its adoption. Regulations adopted by Climate Control could provide exceptions to the prohibition, as well as the

---

[8] "Fossil Fuel Non-Proliferation Treaty," https://fossilfueltreaty.org.
[9] UN Environment Programme and Stockholm Environment Institute, "Production Gap Report 2021" (UNEP, 2021), https://productiongap.org.

necessary framework and definitions for knowing exactly what constitutes net deforestation.

The GCT would also research the best ways to sequester carbon in existing forests, and in agriculture, and require that state parties adopt policies requiring the best-available sequestration. It could also require States that have engaged in deforestation to restore some of their forests.

GCT finance

The sale of GHG emission allowances should provide adequate funding to implement the treaty, by beefing up the UNFCCC secretariat and establishing the Climate Control division, which will need to become a fairly large bureaucracy fairly quickly. The GCT should contain a provision allowing Climate Control to obtain more funds from State parties in equitable proportions (more from rich countries and countries that have high historical emissions) through assessments.

Climate Control would also provide funds and services to developing countries to help them develop their economies and legal systems in climate-friendly ways, including support for mitigation and adaptation, as well as reparations for climate damages inflicted upon them by the major emitters.

Research and development

Besides supporting Climate Control and the UNFCCC Secretariat, revenues would be used to fund important research and development. It would give grants to scientific and R&D organizations to support important climate-related research and development, in areas described above.

Climate policies

State parties to the GCT would also agree to adopt climate-friendly policies. They would immediately end all subsidies for fossil-fuel exploration and development, and prohibit the construction of oil and gas infrastructure such as oil and gas wells and pipelines. They would outlaw the use of natural gas in new development. They would provide subsidies for renovation of older, less energy-efficient buildings, to remove dependence on natural gas, coal, and heating oil and increase their energy efficiency by sealing leaks and increasing insulation. They would require, as oil and gas production are phased out, that existing wells and infrastructure be capped and sealed to prevent methane leaks.

Climate Control could pass regulation adding to these mandatory policies. They would be expected to become more numerous and stringent as time goes on.

What can individuals do?

Individual action will not be sufficient to halt climate change. But there are many things that individuals can do to help. They can reduce their carbon footprint, by switching to all electric energy instead of natural gas, fuel oil, and coal, for their homes. They can install solar panels and batteries at their homes to reduce their use of grid energy. They can drive electric vehicles instead of fossil-fuel cars. They can eat a more plant-based diet. They can fly less on airplanes. They can take public transportation whenever possible. They can switch to more energy-efficient appliances (such as refrigerators and LED lighting) in their homes, and make sure their homes are well insulated and free of air leaks that waste energy. They can minimize waste by recycling and buying products that have small amounts of material waste. They can have fewer children. They can consume less, in general.

Just as important, maybe more so, individuals can pressure governments and corporations for action on the climate. In order to do this, they need to keep up with climate-related news. But this, by itself, does little good. I'm frustrated because I know so many smart, well-educated people who avidly follows the news, but never take any action in response. They're news hobbyists, and if they would spend one-tenth of the time and effort they spend following the news on helping force positive change they would collectively have a huge impact, but they don't.

The most obvious and widespread way to help is by voting, which is very important in democracies. But those of us who can afford the time should participate in our government by doing more than just voting. I like being involved in public affairs; it helps me feel connected to the social and societal world I live in. For many of us, it's easier to do this at the local and state level than the national level. Here are some ideas:

- Pick an issue that seems important to you and follow it in local or state government; organize comments and lobby. For example, if you think it's important to install more EV chargers where they are publicly accessible, then follow what your city, province, or national legislature is doing on this issue. Look for times when your influence can be felt, such as when you can comment, in a public hearing, on a bill or ordinance being considered by a legislative body like the city council.
- Donate money and provide volunteer support to climate causes.
- Discuss climate change with your friends, or small groups you participate in, when you can. You'd be surprised how many people never even give it a thought.
- Those with appropriate professional skills, such as lawyers and scientists, can use those skills in support of a livable climate for this planet. I make a living as a lawyer suing to reduce the GHG emissions of developments, such as warehouses and housing projects. The scientists who contribute to the IPCC reports make a huge contribution by advancing our knowledge of climate science, and how it fits into our societal and economic systems. Many other professions, such as publicists, engineers, city planners, and corporate managers can make big contributions within their disciplines.
- Attend public demonstrations on the climate and related issues.
- Help make climate a more visible issue in politics by participating in local political committees that promote political candidates and set political platforms.
- Demand action on climate from elected officials.
- Divest any investments you might have from institutions that fund fossil-fuel expansion, such as oil companies and banks that finance fossil-fuel projects.

Conclusion

I spend a fair amount of creative energy thinking of ways to fight global heating. As a lawyer I can be most effective using legal means. In particular I try to find novel legal theories that would support lawsuits that would help reduce GHG emissions. You probably have a different skill set, and can help in a different way. An artist could highlight the issues; I'm surprised how few novels and movies have a climate theme. Notable exceptions are *The Ministry for the Future* by Kim Stanley Robinson and the movie *Don't Look Up*. I urge you to consider how you might use your talents and skills to help in this important fight.

Sources

"Ahimsa." In *Wikipedia*. https://en.wikipedia.org/wiki/Ahimsa.

American Electric Power Company, Inc. v. Connecticut, 564 U.S. 410 (2011).

"Arguments from Global Warming Skeptics and What the Science Really Says." In Skeptical Science. https://skepticalscience.com/argument.php.

Arrhenius, Svante. "On the Influence of Carbonic Acid in the Air upon the Temperature of the Ground." *Philosophical Magazine and Journal of Science*, 5, 41 (April 1896): 237–76. https://www.rsc.org/images/Arrhenius1896_tcm18-173546.pdf.

Baker, Jennifer L., Charles N. Rotimi, and Daniel Shriner. "Human Ancestry Correlates with Language and Reveals That Race Is Not an Objective Genomic Classifier." *Scientific Reports* 7 (2017): 1572. https://doi.org/10.1038/s41598-017-01837-7.

Bannock, Graham, and R.E. Baxter. "Theories of Value." In *The Penguin Dictionary of Economics*, 412. London: Penguin Books, 2011.

Bateson, M., S Desire, S.E. Gartside, and G.A. Wright. "Agitated Honeybees Exhibit Pessimistic Cognitive Biases." *Current Biology* 21, no. 12 (2011): 1070–73.

Bayer, Patrick, and Michaël Aklin. "The European Union Emissions Trading System Reduce CO2 Emissions despite Low Prices." *PNAS* 117, no. 16 (April 6, 2020): 8804–12. https://doi.org/10.1073/pnas.191812811.

BBC. "Brazil: Amazon Sees Worst Deforestation Levels in 15 Years." BBC, November 19, 2021. https://www.bbc.com/news/world-latin-america-59341770.

Bible, n.d.

Bloom, Nicholas, Charles I. Jones, John Van Reenen, and Michael Webb. "Are Ideas Getting Harder to Find?" *American Economic Review* 110, no. 4 (2020): 1104–44. https://doi.org/10.1257/aer.20180338.

Bloomberg News. "China Wants More Climate Court Cases, But Only the Right Ones." Bloomberg, June 19, 2021. https://www.bloomberg.com/news/articles/2021-06-19/climate-litigation-must-navigate-china-s-complex-legal-system.

Boonin, David. *The Non-Identity Problem & the Ethics of Future People*. Oxford: Oxford University Press, 2014.

Bouche, Teryn, and Laura Rivard. "America's Hidden History: The Eugenics Movement." *Scitable*, September 18, 2014. https://www.nature.com/scitable/forums/genetics-generation/america-s-hidden-history-the-eugenics-movement-123919444/

Boulton, Chris A., Timothy M. Lenton, and Niklas Boers. "Pronounced Loss of Amazon Rainforest Resilience since the Early 2000s." *Nature Climate Change* 12, no. March 2022 (March 2022): 271–80. https://doi.org/10.1038/s41558-022-01287-8.

Bradley, Curtis A. *International Law in the U.S. Legal System*. Oxford: Oxford University Press, 2013.

Brand, Stewart. *Space Colonies*. Middlesex: Penguin Books, 1977.

Broome, John. *Weighing Lives*. Oxford: Oxford University Press, 2004.

Burke, Marshall, W. Mathew Davis, and Noah S. Diffenbaugh. "Large Potential Reduction in Economic Damages under UN Mitigation Targets." *Nature* 557, no. 7706 (May 24, 2018): 549–53.

Cai, Yongyan, Timothy M. Lenton, and Thomas S. Lontzek. "Risk of Multiple Interacting Tipping Points Should Encourage Rapid CO2 Emission Reduction." *Nature Climate Change* 6, no. May 2016 (March 21, 2016): 520–25. https://doi.org/10.1038/NCLIMATE2964.

California Air Resources Board. "2022 Scoping Plan for Achieving Carbon Neutrality," November 16, 2022. https://ww2.arb.ca.gov/sites/default/files/2022-12/2022-sp.pdf.

Calthorpe, Peter. *Urbanism in the Age of Climate Change*. Washington, DC: Island Press, 2011.

Cao, Yin. "Biodiversity Ruling to Block Dam Project First of Its Kind in China." China Daily, February 21, 2022. https://www.chinadaily.com.cn/a/202202/21/WS62133309a310cdd39bc8 7ef5.html.

Carleton, Tamma A. et al. "Valuing the Global Mortality Consequences of Climate Change Accounting for Adaptation Costs and Benefits." Cambridge, MA: National Bureau of Economic Research, April 2022. https://www.nber.org/system/files/working_papers/w27599/w27599.pdf

"Carrying Capacity." In American Heritage Dictionary. Houghton Mifflin Harcourt, 2020.

"Census Finds U.S. Population Will Decline Without Immigration," Cato Institute, Feb. 14, 2020, https://www.cato.org/blog/census-finds-us-population-will-decline-without-immigration

"China's Achievements, New Goals, and New Measures for Nationally Determined Contributions." China, 2021. https://unfccc.int/sites/default/files/NDC/2022-06/China's%20Achievements%2C%20New%20Goals%20and%20New%20Measures%20for%20Nationally%20Determined%20Contributions.pdf

"Chronological Overview." EUR-Lex, 2020. https://eur-lex.europa.eu/collection/eu-law/treaties/treaties-overview.html.

Cleetus, Rachel. "The Social Cost of Carbon Gets an Interim Update from the Biden Administration." The Equation (blog), March 2, 2021. https://blog.ucsusa.org/rachel-cleetus/the-social-cost-of-carbon-gets-an-interim-update-from-the-biden-administration/.

Climate Case Chart. "Notre Affaire à Tous and Others v. France." Climate Case Chart, 2020. http://climatecasechart.com/non-us-case/notre-affaire-a-tous-and-others-v-france/

———. "Urgenda Foundation v. State of the Netherlands." Climate Case Chart, 2020. http://climatecasechart.com/non-us-case/urgenda-foundation-v-kingdom-of-the-netherlands/

Coghill, Ken, Charles Sampford, and Tim Smith. *Fiduciary Duty and the Atmospheric Trust*. New York: Routledge, 2012.

Cohen, Joel E. *How Many People Can the Earth Support?* New York: W. W. Norton & Company, Inc., 1995.

"Congress Climate History." Center for Climate and Energy Solutions. https://www.c2es.org/content/congress-climate-history/.

Cook, John, Naomi Oreskes, Peter T. Doran, William R. L. Anderegg, Bart Verheggen, Ed W. Maibach, J. Stuart Carlton, et al. "Consensus on Consensus: A Synthesis of Consensus Estimates on Human-Caused Global Warming." *Environmental Research Letters* 11, no. 4 (April 2016): 048002. https://doi.org/10.1088/1748-9326/11/4/048002.

COP21. "Paris Agreement." United Nations, 2015. https://unfccc.int/sites/default/files/english_paris_agreement.pdf.

Cosmides, Leda, and John Tooby. "Evolutionary Psychology: A Primer." UCSB Center for Evolutionary Psychology, 1997. https://www.cep.ucsb.edu/primer.html.

Creutzig, Felix, and et al. "Towards Demand-Side Solutions for Mitigating Climate Change." *Nature Climate Change* 8 (2018): 260–63.

Cyranoski, David. "The CRISPR-Baby Scandal: What's next for Human Gene-Editing." *Nature* 556 (2019): 440–42. https://doi.org/10.1038/d41586-019-00673-1.

Darwin, Charles. *The Formation of Vegetable Mould through the Action of Worms: With Observations on Their Habits*. London: John Murray, 1881.

De Vos, Jurriaan M., and et al. "Estimating the Normal Background Rate of Species Extinction." *Conservation Biology* 29, no. 2 (2014): 452–62.

Dessler, Andres. *Introduction to Modern Climate Change*. 2nd Ed. Cambridge, UK: Cambridge University Press, 2016.

"Domestication of the Dog." In *Wikipedia*. https://en.wikipedia.org/wiki/Domestication_of_the_dog.

Drupp, Moritz A., Mark C. Freeman, Ben Groom, and Frikk Nesje. "Discounting Disentangled." *American Economic Journal: Economic Policy 2018* 10, no. 4 (November 2018): 109–34.

Earthjustice. "What the Inflation Reduction Act Means for Climate," August 16, 2022. https://earthjustice.org/brief/2022/what-the-inflation-reduction-act-means-for-climate.

Editorial. "Why Current Negative-Emissions Strategies Remain 'Magical Thinking.'" *Nature* 554, no. 404 (2018). https://doi.org/10.1038/d41586-018-02184-x.

"Eemian." In *Wikipedia*, https://en.wikipedia.org/wiki/Eemian.

Eldredge, Niles. "A Field Guide to the Sixth Extinction." *New York Times*, 1999. https://archive.nytimes.com/www.nytimes.com/library/magazine/millennium/m6/extinction-eldredge.html.

Elkins, Zachary, Tom Ginsburg, and James Melton. *Conceptualizing Constitutions: The Endurance of National Constitutions*. Cambridge, UK: Cambridge University Press, 2009.

"Emissions Gap Report 2022." UNEP, 2022. https://www.unep.org/resources/emissions-gap-report-2022.

"Environment." In *Oxford English Dictionary*, 1971.

Environmental Pollution Panel, President's Science Advisory Committee. "Restoring the Quality of Our Environment. Report," November 1965. https://legacy-assets.eenews.net/open_files/assets/2019/01/11/document_cw_01.pdf.

Esipova, Neli, Anita Pugliese, and Julie Ray. "More than 750 Million Worldwide Would Migrate If They Could." Gallup, December 10, 2018. https://news.gallup.com/poll/245255/750-million-worldwide-migrate.aspx.

EUR-Lex. "Ratification Process." European Union. https://eur-lex.europa.eu/EN/legal-content/glossary/ratification-process.html.

"European Climate Law." European Commission, 2021. https://climate.ec.europa.eu/eu-action/european-green-deal/european-climate-law_en.

European Commission. "EU Emissions Trading System (EU ETS)."
    https://climate.ec.europa.eu/eu-action/eu-emissions-trading-system-
    eu-ets_en.

— — —. "Update of the NDC of the European Union and Its Member
    States," December 17, 2020.
    https://unfccc.int/sites/default/files/NDC/2022-
    06/EU_NDC_Submission_December%202020.pdf.

Evans, Simon. "Which Countries Are Historically Responsible for Climate
    Change?" Carbon Brief, May 10, 2021.
    https://www.carbonbrief.org/analysis-which-countries-are-historically-
    responsible-for-climate-change/.

"Fermi Paradox." In *Wikipedia*.
    https://en.wikipedia.org/wiki/Fermi_paradox.

"Fossil Fuel Non-Proliferation Treaty." https://fossilfueltreaty.org.

Gardiner, Stephen M. *A Perfect Moral Storm: The Ethical Tragedy of Climate
    Change*. Oxford: Oxford University Press, 2011.

Geer v. State of Conn., 161 U.S. 519 (U.S. 1896).

Green Climate Fund. "Status of Pledges and Contributions Made to the
    Green Climate Fund." Green Climate Fund, July 31, 2020.
    https://www.greenclimate.fund/sites/default/files/document/status-
    pledges-irm_1.pdf.

Green, Mark. "Chinese Coal-Based Power Plants." Wilson Center.
    https://www.wilsoncenter.org/blog-post/chinese-coal-based-power-
    plants.

Greenpeace International. "Saudi Arabian Negotiators Move to Cripple
    COP26 – Greenpeace Response." Greenpeace International, November
    7, 2021.
    https://www.greenpeace.org/international/press-release/50547/cop26-
    saudi-arabia-negotiators-cripple/.

Gumprecht, Blake. *The Los Angeles River*. Baltimore: The Johns Hopkins
    University Press, 1999.

Gye, Hugo. "America IS the 1%: You Need Just $34,000 Annual Income to Be in the Global Elite... and HALF the World's Richest People Live in the U.S." Daily Mail, January 5, 2012. https://www.dailymail.co.uk/news/article-2082385/We-1--You-need-34k-income-global-elite--half-worlds-richest-live-U-S.html.

Hallam, A., and P.B. Wignall. *Mass Extinctions and Their Aftermath*. Oxford: Oxford University Press, 1997.

Harari, Yuval Noah. *Sapiens: A Brief History of Humankind*. Signal Books, 2014.

Hardin, Garrett. "The Tragedy of the Commons." *Science* 162, no. 3859 (December 13, 1968): 1243–48. https://doi.org/10.1126/science.162.3859.1243.

Hellman, Martin E., and Vinton G. Cerf. "An Existential Discussion: What Is the Probability of Nuclear War?" *Bulletin of the Atomic Scientists*, March 18, 2021. https://thebulletin.org/2021/03/an-existential-discussion-what-is-the-probability-of-nuclear-war/#post-heading.

Houghton, John. *Global Warming - The Complete Briefing*. 5th ed. Cambridge, UK: Cambridge University Press, 2015.

"How Is China Tackling Climate Change?" Grantham Research Institute on Climate Change and the Environment, 97/25/22. https://www.lse.ac.uk/granthaminstitute/explainers/how-is-china-tackling-climate-change/.

"Human Rights." United Nations. https://www.un.org/en/global-issues/human-rights.

Illinois Central Railroad Co. v. Illinois, 146 U.S. 387 (1892).

"Industrial Revolution." In *Wikipedia*, September 6, 2021. https://en.wikipedia.org/w/index.php?title=Industrial_Revolution&oldid=1042697673.

International Energy Agency. "Energy Efficiency 2020." https://www.iea.org/reports/energy-efficiency-2020.

International Telecommunications Union. "Facts and Figures 2021: 2.9 Billion People Still Offline." ITU, November 29, 2021. https://www.itu.int/hub/2021/11/facts-and-figures-2021-2-9-billion-people-still-offline/.

IPCC. "AR5 Synthesis Report: Climate Change 2014," 2015. https://www.ipcc.ch/report/ar5/syr/.

———. "Climate Change and Land." IPCC, August 7, 2019. https://www.ipcc.ch/srccl/.

IPCC WGI. "Climate Change 2021: The Physical Science Basis." IPCC WGI, 2021. https://www.ipcc.ch/report/ar6/wg1/.

IPCC WGII. "Climate Change 2022: Impacts, Adaptation and Vulnerability." IPCC, 2022. https://report.ipcc.ch/ar6/wg2/IPCC_AR6_WGII_FullReport.pdf.

IPCC WGIII. "Climate Change 2022: Mitigation of Climate Change." IPCC, 2022. https://www.ipcc.ch/report/ar6/wg3/.

James Hansen. *Storms of My Grandchildren*. Bloomsbury USA, 2009.

Judt, Tony. *Postwar: A History of Europe since 1945*. Penguin Books, 2005.

Kerschner, Seth. "United States: Greenhous Gas Emissions Trading Schemes." Lexology/White & Case LLP, September 20, 2017. https://www.lexology.com/library/detail.aspx?g=0f6bf054-27dd-4cc0-b856-107b1ad0854e.

Klein, Naomi. *This Changes Everything: Capitalism vs the Climate*. New York: Simon & Schuster, 2014.

Knox, John. "Constructing the Human Right to a Healthy Environment." Annual Review of Law and Social Science, July 27, 2020. https://doi.org/10.1146/annurev-lawsocsci-031720-07-4856.

Kolbert, Elizabeth. *The Sixth Extinction*. Henry Holt, 2014.

"Kurd People." In *Encyclopedia Britannica*, November 28, 2022. https://www.britannica.com/topic/Kurd.

Lane, Lea. "Percentage Of Americans Who Never Traveled Beyond The State Where They Were Born? A Surprise." *Forbes*, May 2, 2019.

https://www.forbes.com/sites/lealane/2019/05/02/percentage-of-americans-who-never-traveled-beyond-the-state-where-they-were-born-a-surprise/?sh=203e48792898.

"Last Glacial Period." In *Wikipedia*, n.d. https://en.wikipedia.org/wiki/Last_Glacial_Period.

Lees, Emma, and Jorge E. Viñuales. *The Oxford Handbook of Comparative Environmental Law*. Oxford: Oxford University Press, 2019.

Leopold, Aldo. *Sand Country Almanac, Special Commemorative Edition*. Oxford: Oxford University Press, 1987.

Leopold, Luna B. *Waters, Rivers and Creeks*. Sausalito, CA: University Science Books, 1997.

"Liberalism" in *American Heritage Dictionary*. 5th ed. Boston: Houghton-Mifflin, 2011.

"List of Nationalizations by Country." In *Wikipedia*. https://en.wikipedia.org/wiki/List_of_nationalizations_by_country.

Liu, Peiran R., and Adrian E. Rafferty. "Country-Based Rate of Emissions Reductions Should Increase by 80% beyond Nationally Determined Contributions to Meet the 2°C Target." *Nature Communications Earth & Environment* 2, no. 29 (February 9, 2021). https://doi.org/10.1038/s43247-021-00097-8.

"Living Planet Report 2022." World Wildlife Fund, October 13, 2022. https://livingplanet.panda.org/en-US/.

"Lochner Era" in *Wikipedia*, https://en.wikipedia.org/wiki/Lochner_era.

Lopez, Barry. "The Passing Wisdom of Birds." In *Crossing Open Ground*. New York: Vintage, 1989.

Lord, N.S., A. Ridgwell, M.C. Thorne, and D.J. Lunt. "An Impulse Response Function for the 'Long Tail' of Excess Atmospheric CO2 in an Earth System Model." *Global Biogeochem Cycles* 30 (January 9, 2016): 2–17. https://doi.org/10.1002/2014GB005074.

Lovelock, James. *The Vanishing Face of Gaia: A Final Warning*. Basic Books, 2009.

Lynas, Mark. *Our Final Warning: Six Degrees of Climate Emergency*. London: 4th Estate, 2020.

MacAskill, William. *What We Owe the Future*. New York: Basic Books, 2022.

Mack, Chris. "The Multiple Lives of Moore's Law." *IEEE Spectrum*, March 30, 2015. https://spectrum.ieee.org/the-multiple-lives-of-moores-lawwww.mooreslaw.org.

"Maslow's Hierarchy of Needs." In *Wikipedia*. https://en.wikipedia.org/wiki/Maslow%27s_hierarchy_of_needs.

Mathez, Edmon A., and Jason E. Smerdon. *Climate Change: The Science of Global Warming*. 2nd ed. New York: Columbia University Press, 2018.

McConnell, Campbel R., Stanley L. Brue, and Sean M. Flynn. *Macroeconomics*. 21st ed. New York: McGraw-Hill Education, 2018.

McKibben, Bill. *Falter: Has the Human Game Begun to Play Itself Out?* New York: Holt Paperbacks, 2019.

Miller, Sabbie A., and et al. "Achieving Net Zero Greenhouse Gas Emissions in the Cement Industry via Value Chain Mitigation Strategies." *One Earth* 4, no. 10 (October 22, 2021): 1398–1411. https://doi.org/10.1016/j.oneear.2021.09.011.

Mora, Camilo et al. "Global Risk of Deadly Heat." *Nature Climate Change* 7, no. July 2017 (June 19, 2017): 501–7. https://doi.org/10.1038/NCLIMATE3322.

Morris, Simon Conway. *The Runes of Evolution: How the Universe Became Self-Aware*. West Conshohocken, PA: Templeton Press, 2015.

"Mortality Risk Valuation." US Environmental Protection Agency, March 30, 2022. https://www.epa.gov/environmental-economics/mortality-risk-valuation.

Murphy, Sean D. *Principles of International Law*. St. Paul, MN: Thomson/West, 2006.

Naess, Arne. "The Shallow and the Deep, Long-range Ecology Movement. A Summary." *Inquiry* 16, no. 1–4 (August 29, 2008): 95–100.

NASA Global Climate Change. "Global Surface Temperature | NASA Global Climate Change." Climate Change: Vital Signs of the Planet. https://climate.nasa.gov/vital-signs/global-temperature.

National Audubon Society v. Superior Court, 33 Cal.3d 419 (Cal. 1983).

National Research Council. "Climate Intervention: Carbon Dioxide Removal and Reliable Sequestration." National Academies Press, 2015. http://nap.nationalacademies.org/18805.

Nordhaus, William. *The Climate Casino: Risk, Uncertainty, and Economics for a Warming World*. New Haven: Yale University Press, 2013.

O' Neill, Daniel W., and et al. "A Good Life for All within Planetary Boundaries." *Nature Sustainability* 1 (February 2018): 88–95.

OECD. "Climate Finance and the USD 100 Billion Goal." OECD, September 22, 2022. https://www.oecd.org/climate-change/finance-usd-100-billion-goal/.

———. "Labour Productivity and Utilisation." OECD, 2021. https://data.oecd.org/lprdty/labour-productivity-and-utilisation.htm#indicator-chart.

Ord, Toby. *The Precipice: Existential Risk and the Future of Humanity*. New York: Hachette, 2020.

Our World in Data. "Who Has Contributed Most to Global CO2 Emissions?," 2019. https://ourworldindata.org/contributed-most-global-co2.

Parfit, Derek. *Reasons and Persons*. Oxford: Oxford University Press, 1984.

Polar Portal. "Mass and Height Change, Greenland Ice Sheet." http://polarportal.dk/en/greenland/mass-and-height-change/.

Poore, Richard, Richard S. Williams, and Christopher Tracey. "Sea Level and Climate." USGS, n.d. https://pubs.usgs.gov/fs/fs2-00/.

Popp, Max, Hauke Schmidt, and Joche Marotzke. "Transition to a Moist Greenhouse with CO2 and Solar Forcing." *Nature Communications* 7, no. 10627 (February 9, 2016). https://doi.org/10.1038/ncomms10627.

"Poverty." World Bank, November 30, 2022. https://www.worldbank.org/en/topic/poverty/overview.

Princeton University Zero Lab. "Preliminary Report: The Climate and Energy Impacts of the Inflation Reduction Act of 2022," August 2022. https://repeatproject.org/docs/REPEAT_IRA_Prelminary_Report_2022-08-04.pdf.

"Progressive Era" In *Wikipedia*, https://en.wikipedia.org/wiki/Progressive_Era

"Public Nuisance." In *Black's Law Dictionary*. St. Paul, MN: Thompson Reuters, 2019.

Ramanathan, V., and Y Feng. "On Avoiding Dangerous Anthropogenic Interference with the Climate System: Formidable Challenges Ahead." *Proceedings of the National Academy of Sciences* 105, no. 38 (2008): 14245–50.

Raworth, Philip. *Introduction to the Legal System of the European Union*. Dobbs Ferry: Oceana Publications, 2001.

"Responding to Climate Change: China's Policies and Actions." State Council Information Office of the People's Republic of China, October 27, 2021. http://www.scio.gov.cn/zfbps/32832/Document/1715506/1715506.htm.

Rice, D.P., and B.S. Cooper. "The Economic Value of Human Life." *Am J Public Health Nations Health* 57, no. 11 (1967). https://pubmed.ncbi.nlm.nih.gov/6069745/.

Ricke, Katharine, Laurent Drouet, Ken Caldeira, and Massimo Tavoni. "Country-Level Social Cost of Carbon." *Nature Climate Change* 8, no. October 2018 (2018): 895–900.

"Right." In *Black's Law Dictionary*. St. Paul, MN: West Publishing Co., 2019.

Roberts, J. Timmons, Romain Weikmans, Stacy-ann Robinson, David Ciplet, Mizan Khan, and Danielle Falzon. "Rebooting a Failed Promise of Climate Finance." *Nature Climate Change* 11 (February 18, 2021): 180–82. https://www.nature.com/articles/s41558-021-00990-2.

Robertson, Margaret. *Sustainability: Principles and Practice*. 3d. New York: Routledge, 2021.

Rosenfeld, Michel, and András Sajó. *The Oxford Handbook of Comparative Constitutional Law*. Oxford: Oxford University Press, 2012.

Sax, Joseph. "The Public Trust Doctrine in Natural Resource Law: Effective Judicial Intervention." *Mich. L. Rev.* 68 (1970): 471–566. https://repository.law.umich.edu/mlr/vol68/iss3/3.

Schneider, Tapio, Colleen M. Kaul, and Kyle G. Pressel. "Possible Climate Transitions from Breakup of Stratocumulus Decks under Greenhouse Warming." *Nature Geoscience* 12 (March 2019): 163–67. https://doi.org/10.1038/s41561-019-0310-1.

Schweiger, Axel, Ron Lindsay, Jinlun Zhang, Mike Steele, Harry Stern, and Ron Kwok. "Uncertainty in Modeled Arctic Sea Ice Volume." *Journal of Geophysical Research* 116 (September 27, 2011): C00D06. https://doi.org/10.1029/2011JC007084.

Schweitzer, Albert. *Civilization and Ethics.* 3rd ed. London:Adam & Charles Black, 1961.

Science Daily. "Evolution Imposes 'speed Limit' on Recovery after Mass Extinctions," April 8, 2019. https://www.sciencedaily.com/releases/2019/04/190408114252.htm.

ScienceDirect. "Life Span." *ScienceDirect* (blog), 2013. https://www.sciencedirect.com/topics/earth-and-planetary-sciences/life-span.

"Scientific Method." In *Wikipedia*, https://en.wikipedia.org/w/index.php?title=Scientific_method&oldid=1044623103.

Smil, Vaclav. "A Global Transition to Renewable Energy Will Take Many Decades." *Scientific American,* January 2014. https://www.scientificamerican.com/article/a-global-transition-to-renewable-energy-will-take-many-decades/.

Smith, Adam. *Wealth of Nations.* London: Penguin Books, 1776.

Smith, Rhona K.M. *International Human Rights Law.* 10th ed. Oxford: Oxford University Press, 2022.

Smith-Heimer, Janet, Jessica Hitchcock, and Greg Goodfellow. "CEQA: California's Living Environmental Law." Berkeley, CA: The Housing Workshop, October 2021.

https://rosefdn.org/wp-content/uploads/CEQA-California_s-Living-Environmental-Law-10-25-21.pdf.

Smith, Zadie. "Fascinated to Presume: In Defense of Fiction," The New York Review of Books, Oct. 24, 2019.

"Special Report on the Ocean and Cryosphere in a Changing Climate." IPCC, 2019. https://www.ipcc.ch/srocc/.

Srinivasan, Balaji. *The Network State: How to Start a New Country*, 2022. https://thenetworkstate.com.

Stern, Nicholas. *The Economics of Climate Change: The Stern Review*. Cambridge, UK: Cambridge University Press, 2006.

Stevens, Mary. "The Precautionary Principle in the International Arena." *Sustainable Development Law & Policy* 2, no. 2 (2002): 6. https://digitalcommons.wcl.american.edu/cgi/viewcontent.cgi?referer=&httpsredir=1&article=1278&context=sdlp.

Subramanian, Meera. "Anthropocene Now: Influential Panel Votes to Recognize Earth's New Epoch." *Nature*, May 21, 2019. https://www.nature.com/articles/d41586-019-01641-5.

Swiss Re Institute. "The Economics of Climate Change: No Action Not an Option." Swiss Re Institute, April 2021. https://www.swissre.com/institute/research/topics-and-risk-dialogues/climate-and-natural-catastrophe-risk/expertise-publication-economics-of-climate-change.html.

Tabuchi, Hiroko. "Inside the Saudi Strategy to Keep the World Hooked on Oil." *New York Times*, November 21, 2022. https://www.nytimes.com/2022/11/21/climate/saudi-arabia-aramco-oil-solar-climate.html.

Takakura, Jun'ya, and et al. "Dependence of Economic Impacts of Climate Change on Anthropogenically Directed Path." *Nature Climate Change* 9 (October 2019): 737–41. https://doi.org/10.1038/s41558-019--578-6.

"The United States of America Nationally Determined Contribution," April 2021.

https://unfccc.int/sites/default/files/NDC/2022-06/United%20States%20NDC%20April%2021%202021%20Final.pdf.

The World Bank. "Trends in Solid Waste Management." World Bank. https://datatopics.worldbank.org/what-a-waste/trends_in_solid_waste_management.html.

Thevenon, Gilles. *Histoire des Constitutions: Vie politique française 1789 à 1958*. Lyon: Chronique Sociale, 2017.

Treaty establishing the European Coal and Steel Community (1951). https://eur-lex.europa.eu/legal-content/EN/LSU/?uri=CELEX:11951K/TXT.

Treaty establishing the European Economic Community (1957). https://eur-lex.europa.eu/EN/legal-content/summary/treaty-of-rome-eec.html.

Treaty on European Union (1992). https://eur-lex.europa.eu/legal-content/EN/TXT/PDF/?uri=CELEX:11992M/TXT&from=EN.

Treaty on the Functioning of the European Union (2007). https://eur-lex.europa.eu/EN/legal-content/summary/treaty-on-the-functioning-of-the-european-union.html.

"Treaty Ratification." American Civil Liberties Union, 2022. https://www.aclu.org/issues/human-rights/treaty-ratification.

Tyrrell, Toby. *On Gaia: A Critical Investigation of the Relationship between Life and Earth*. Princeton, NJ: Princeton University Press, 2013.

UN Environment Programme, and SEI, IISD, ODI, E3G, and UNEP. "Production Gap Report 2021." UNEP, 2021. https://productiongap.org/2021report.

UN Food and Agriculture Organization. "The State of the World's Forests." UN FAO, 2020. https://www.fao.org/3/ca8642en/ca8642en.pdf.

"U.N. General Assembly Resolution 2542, Declaration on Social Progress and Development." United Nations, December 11, 1969. https://www.ohchr.org/sites/default/files/Documents/ProfessionalInterest/progress.pdf.

UN Human Rights Council. "Right to a Healthy Environment: Good
    Practices." United Nations Human Rights Council, 2020 209AD.
    https://documents-dds-
    ny.un.org/doc/UNDOC/GEN/G19/355/14/PDF/G1935514.pdf?OpenEle
    ment.

UN News. "UN General Assembly Declares Access to Clean and Healthy
    Environment a Universal Human Right." New York: United Nations,
    July 28, 2022. https://news.un.org/en/story/2022/07/1123482.

UN World commission on Environment and Development. "Report of the
    World Commission on Environment and Development: Our Common
    Future," 1987.
    https://sustainabledevelopment.un.org/content/documents/5987our-
    common-future.pdf.

UNEP. "What Is the United Nations Framework Convention on Climate
    Change?" United Nations.
    https://unfccc.int/process-and-meetings/what-is-the-united-nations-
    framework-convention-on-climate-change.

United Nations. "2030 Agenda for Sustainable Development," September
    27, 2015. https://sdgs.un.org/2030agenda.

— — —. "Goal 12: Ensure Sustainable Consumption and Production
    Patterns." Sustainable Development Goals. United Nations.
    https://www.un.org/sustainabledevelopment/sustainable-
    consumption-production/.

— — —. "Report of the Conference of the Parties on Its Twenty-First Session,
    Held in Paris from 30 November to 13 December 2015." United Nations,
    January 29, 2016.
    https://unfccc.int/resource/docs/2015/cop21/eng/10a01.pdf.

— — —. "U.N. Sustainable Development Goals," 2015.
    https://sdgs.un.org/goals.

United Nations Dept. of Economic and Social Affairs, Population Division.
    "World Population Prospects 2022: Summary of Results," 2022.
    https://www.un.org/development/desa/pd/sites/www.un.org.develop
    ment.desa.pd/files/wpp2022_summary_of_results.pdf.

"United States Ratification of International Human Rights Treaties." Human Rights Watch, July 24, 2009. https://www.hrw.org/news/2009/07/24/united-states-ratification-international-human-rights-treaties.

U.S. EPA. "Inventory of U.S. Greenhouse Gas Emissions and Sinks," April 4, 2022. https://www.epa.gov/ghgemissions/inventory-us-greenhouse-gas-emissions-and-sinks.

Wahal, Anya. "On International Treaties, the United States Refuses to Play Ball." Council on Foreign Relations, January 7, 2022. https://www.cfr.org/blog/international-treaties-united-states-refuses-play-ball.

Ward, James D., Paul C. Sutton, Arian D. Werner, Robert Costanza, Steve H. Morh, and Craig T. Simmons. "Is Decoupling GDP Growth from Environmental Impact Possible?" *Plos ONE* 11, no. 10 (2016). https://journals.plos.org/plosone/article/file?id=10.1371/journal.pone.0164733&type=printable.

Ward, Peter. "What May Become of Homo Sapiens." *Scientific American*, November 1, 2012. https://www.scientificamerican.com/article/what-may-become-of-homo-sapiens/#.

"Well-Being Concepts." Centers for Disease Control and Prevention, October 31, 2018. https://www.cdc.gov/hrqol/wellbeing.htm.

World Bank. "CO2 Emissions." World Bank, 2020. https://data.worldbank.org/indicator/EN.ATM.CO2E.PC.

— — —. "GDP Growth," 2022. https://data.worldbank.org/indicator/NY.GDP.MKTP.KD.ZG.

World Health Organization. "9 out of 10 People Worldwide Breathe Polluted Air, but More Countries Are Taking Action." World Health Organization, May 2, 2018. https://www.who.int/news-room/detail/02-05-2018-9-out-of-10-people-worldwide-breathe-polluted-air-but-more-countries-are-taking-action.

"World Migration Report 2022." International Organization for Migration (IOM), 2021. https://publications.iom.int/books/world-migration-report-2022.

Acknowledgements

My wife of 48 years, Benita, has been a constant support and help through the vicissitudes of my life, and has encouraged me to take the big step of writing this, my first book. She also provided many valuable suggestions for improving it.

Aria Soeprono, my firm's law clerk, read the manuscript of this book and provided a great deal of insightful and valuable feedback.

I'm proud to be part of a community of plaintiff's-side lawyers and activists in California that fight for the environment. They have supported me in my relatively late environmental law career, and, for that, I'm grateful to them, especially Douglas Carstens.

I also thank Dan Selmi, my excellent professor of environmental law at Loyola Law School, Los Angeles, for his teaching and help in becoming a good lawyer.

I'm grateful to Ethics Press for affording me the opportunity of publishing this, my first book, and guiding me through the process.

And I must acknowledge my parents, Charles F. Wallraff and Evelyn B. Wallraff. My father was a professor of philosophy at the University of Arizona, and my mother was a PhD microbiologist who taught at the university level and also conducted research on Valley Fever. They provided a supportive home and encouraged me to explore as many areas of knowledge as possible.

www.ingramcontent.com/pod-product-compliance
Lightning Source LLC
Chambersburg PA
CBHW021435180326
41458CB00001B/288